Library of
Davidson College

# THE AIR AND SEA LANES OF THE NORTH ATLANTIC: THEIR SECURITY IN THE 1980'S

Sherwood S. Cordier

University Press of America

Copyright © 1981 by
**University Press of America, Inc.**™
P.O. Box 19101, Washington, D.C. 20036

All rights reserved

Printed in the United States of America

ISBN: 0-8191-1587-8

Library of Congress Number: 81-40180

355.03
C 795a

82-8862

## ACKNOWLEDGMENTS

A faculty research fellowship and grant from Western Michigan University helped make this study possible. Valuable assistance was provided by the Royal United Services Institute for Defense Studies in London and its Director, Rear Admiral E.F. Gueritz. Roy Braybrook, Neville Brown, Brian G.R. Thompson, Desmond Wettern, and Derek Wood have been very generous and helpful. Only the author, however, is responsible for the views expressed in this study.

The efforts of Mrs. Judith Massie, who typed the manuscript, are warmly appreciated. I am very grateful to my wife, Mary, for her support and encouragement.

CONTENTS

| | |
|---|---|
| Introduction | 1 |
| Chapter One - The Air Defense of Great Britain | 6 |
| Chapter Two - Challenge on the High Seas: The Soviet Navy | 23 |
| Chapter Three - Control of the Air at Sea | 39 |
| Chapter Four - Confronting the Soviet Submarine Threat | 53 |
| Conclusion | 80 |

INTRODUCTION

The sea and air lanes across the North Atlantic represent the essential lines of communication and transport linking the transatlantic community together. Seaborne trade is immensely important to the countries bordering the North Atlantic. At any given moment some 5,000 western European, British, Canadian, and American merchant vessels ply the waters of the Atlantic. Indeed more than half the shipping of the world sails the trade routes of the Atlantic as a whole.[1] The United States, Great Britain, and such European nations as the Federal Republic of Germany, France, and the Netherlands are major exporters of goods to overseas markets, particularly each other. Moreover, western European nations and the United States are increasingly dependent upon importing vital raw materials from overseas sources. Half of the oil consumed by the United States and ninety percent of the oil needs of western Europe are met by seaborne imports which move into Atlantic ports.[2] Passenger traffic, once carried by proud ocean liners, now moves in a constant stream of aircraft. Thus is the transatlantic community knit together by a vast web of maritime and airborne links.

Above all, the security of western Europe hinges upon the untrammeled use of the North Atlantic. In time of acute crisis or conflict in Europe American troop reinforcements would need to be flown to Europe from the continental United States. American forces stationed in West Germany would be swiftly reinforced by selected regular army units now based in the United States. Such formations include two mechanized infantry divisions, one armored division, and an armored cavalry regiment. Many individual Reserve soldiers and Reserve and National Guard units are also earmarked for duty in the European theatre. Heavy weaponry and equipment for these formations are stockpiled in West Germany. Each year many of these troops are flown to West Germany where they participate in realistic maneuvers in the field. As many as one and a half million American soldiers would be brought into the European arena.[3] Command of the air over western Europe and effective defense of airports are clearly necessary if such massive reinforcement by air is to be successful.

Western European air forces and American air units stationed in Europe will be buttressed by the arrival of some sixty tactical fighter squadrons, about 1,000 warplanes, from the United States. Again, command of the air in the European theatre is absolutely essential if these squadrons are to be received at European airbases and readied to fly into battle.

To sustain American forces waging a desperate struggle in Europe the United States must dispatch across the Atlantic some ten million tons of military supplies and fifteen million tons of fuel within the first ninety days of conflict.[4] And the flow of food, raw materials, petroleum, and key manufactured goods needed by the western European economies must not be interrupted. The overwhelming bulk of these goods and materials can only move by sea. Hence the battle for control of the high seas is crucial.

In the course of the 1970s Soviet tactical airpower

has undergone a sweeping transformation. The capabilities of Soviet tactical squadrons have been immensely enhanced by much more powerful engines, the employment of "swing-wing" technology, new electronic systems, and new warplanes that feature greatly increased payloads and range. The Soviet Union can now unleash savage air raids against seaports, airbases, supply depots, and other vital installations throughout western Europe and the British isles. A powerful and versatile Soviet tactical air force now challenges the western alliance. Confronted by such a foe the United States and its allies will be hard pressed indeed to win the control of the air so crucial to the defense of western Europe.

The Soviet Union is engaged in a massive program dedicated to the development of a major navy. Soviet fleet units are now capable of operating effectively on the high seas. Analyzing Russian maneuvers, Donald Daniel emphasizes:

> If one had to choose an overall qualitative trend from analysis of the exercises, it would be that of an improved capability to operate large numbers of forces in the open ocean, some at very great distances from the Soviet Union.[5]

The most powerful of Soviet surface fleets is based at a vast naval complex on the Kola peninsula near Murmansk. It is thus in a strategic position to move through the Barents and Norwegian Seas upon the North Atlantic. The Russian Northern fleet comprises 32 cruisers and destroyers, 70 frigates and escorts, and 300 smaller surface vessels.[6]

The Soviet navy has amassed a formidable submarine force. Roughly three-fourths of the total Russian submarine fleet, almost 200 of the deadly undersea warships, are based on the Kola peninsula.[7] Under these circumstances it is possible that as many as 600 western merchant vessels and naval escorts could fall victim to enemy submarines in the early stages of any major conflict.[8]

Nor is the threat to the sealanes limited to the submarine. The Soviet Naval Air Force has been substantially expanded and furnished with new aircraft. A maritime attack version of a new fast medium bomber is coming into service with naval squadrons. "Backfire," as this machine is designated in the west, is equipped with new anti-shipping missiles. Flying from airbases in Russia, "Backfire" can range widely over the sea lanes of the Atlantic from Newfoundland to the Azore islands.

How can the western alliance cope with these burgeoning Soviet threats? What measures can the members of the transatlantic community take to assure the security of their vital oceanic links in the 1980s? The Atlantic allies must assure the security of these air and sea lanes in time of peace as well as war. Indeed, in order to deter such conflict, the western allies must be perceived in peacetime to be capable of fulfilling this crucial task successfully. And perceived vulnerability may also invite political and economic blackmail.

Protection of the North Atlantic lines of communication and transport can only be achieved through the close cooperation of the western alliance. Such a common venture clearly underlines the growing integration of the Transatlantic community.

¹Paul H. Nitze, Leonard Sullivan, Jr., and the Atlantic Council Working Group on Securing the Seas, Securing the Seas: The Soviet Naval Challenge and Western Alliance Options (Boulder, Colorado: Westview Press, 1979), p. 145.

²Ibid., p. 122.

³Rear Admiral Sayre A. Swarztrauber, "The Potential Battle of the Atlantic," in Frank Uhlig, Jr., ed., Naval Review 1979 (Annapolis, Maryland: U.S. Naval Institute, 1979), p. 115.

⁴Nitze, et. al., op. cit., p. 380.

⁵Donald C. Daniel, "Trends and Patterns in Major Soviet Naval Exercises," in Paul J. Murphy, ed., Naval Power in Soviet Policy, Studies In Communist Affairs, Volume 2, published under the auspices of the United States Air Force (Washington, D.C.: Government Printing Office, 1978), p. 230.

⁶R.D.M. Furlong, "The Threat to Northern Europe," International Defense Review (No. 4, 1979), p. 523.

⁷Ibid.

⁸Nitze, et. al., op. cit., p. 374.

CHAPTER I
The Air Defense Of Great Britain

The conflict situation in the North Atlantic and western Europe is three dimensional. Any struggle will be waged in the air, by surface warship, and undersea against the submarine. The outcome of a major war, however, will be most decisively shaped by the conflict in the air. This is true for many reasons. Should western airpower be struck down at the very outset of conflict the swift arrival of American warplane squadrons and troop reinforcements would be impossible. Soviet aircraft could range freely over seaports and naval bases and unleash a deadly hail of missiles upon western warships and convoys. The British isles could be isolated, reduced to impotence, and ravaged by air attack.

Frontal Aviation, the tactical branch of Soviet airpower most relevant to air operations in the European arena, has undergone a massive transformation in equipment and capabilities during the course of the 1970s. As the International Institute for Strategic Studies emphasizes:

> It has become clear that the Soviet Union has moved away from the ideal that an air force is an extended arm of the artillery, directly supporting the land battle, to a more Western view of air power as a crucial element in its own right in both tactical and strategic operations. As a consequence she has introduced, and continues to deploy, more complex aircraft designed to deliver heavier weapon loads over longer ranges and with greater accuracy than ever before.[1]

For a number of years the Soviet Union has achieved

a production rate of 1,150 fighters and fighter bombers each year.[2] It has thus been possible for the Soviets to expand their tactical air arm substantially and to replace all older machines with new warplanes. Russian tactical regiments have been much increased in strength, from 36 to 45 aircraft each.[3]

Even more significant is the sharp improvement in Soviet technological capabilities. Quantity is now being matched with quality. New and more powerful turbofan engines have been developed. Variable geometry, the changing of the position of the wings while in flight, is a standard feature of the new Soviet warplanes. New machine cannon and air to air missiles have been introduced. Enormous strides have been made in precision guided air to ground missiles. Such rockets are now standard armament for Soviet strike aircraft.[4]

Most striking of all are continued Soviet advances in sophisticated avionic and electronic equipment.[5] New Soviet planes are now furnished with doppler navigation systems, laser target designators, terrain avoidance radar, multi-mode attack radar, and comprehensive electronic countermeasures suites.

The recently developed Soviet ability to strike deep into western Europe and Britain is well represented in three key new combat aircraft. The MiG-23MF and MiG-27 are companion "swingwing" warplanes that have become the mainstay of Soviet Frontal Aviation.[6] A fighter version, the MiG-23MF is swift and powerfully armed with a 23 mm. machine cannon and four air to air missiles.

Designed for the strike role, the MiG-27 is configured for very low level flight and is equipped with a doppler navigation system, terrain avoidance radar, and a laser target designator. Armed with a six barrel 23 mm. "gatling" machine cannon and four air to surface rockets, the MiG-27 can assail targets as far as southeastern England.

A single engine machine, the MiG-27 lacks the safety margin ensured by twin engine warplanes. Moreover, the work load on the single crew member, who must pilot, navigate, and handle all weapon systems, must be taxing indeed. The International Institute for Strategic Studies indicates that 1,300 MiG-23s and 27s were in service with the Soviet tactical air force in 1980.[7]

Far more formidable is the Sukhoi-24 or Fencer as it is code named in the west. A twin engine "swingwing" with a crew of two, Fencer was designed from its inception as a long range strike and interdiction weapon.

The Sukhoi-24 features a doppler navigation system, a laser target designator, and an array of electronic warfare and countermeasures gear. Particularly significant is the fact that Fencer is the first Soviet tactical warplane to be equipped with a multi-mode attack radar. Flying from bases in eastern Europe, and operating at low level throughout its mission, Fencer can range over the eastern half of England, unleashing more than three tons of explosives.[8] The Sukhoi attack machine is credited with a maximum payload of five tons and can attain a maximum speed at altitudes of at least 1,320 miles per hour.[9]

It has undoubtedly been quite a challenge for the Russians to bring the highly sophisticated Fencer to operational status. Development of the multi-mode radar in particular, proved to be a protracted ordeal.  In 1980 some 370 Sukhoi-24s were in service with Soviet Frontal Aviation.[10]

The Tupolev-22M/26, code named Backfire by NATO, also plays a vital part in the duel for air supremacy in the west.  A medium range bomber and maritime strike warplane, the Backfire serves in approximately equal numbers with Soviet Long Range Aviation and Soviet Naval Aviation. It features powerful twin turbofan engines and a semi-variable geometry design, the "swingwing" pivoting at one-third of the span.  Backfire can reach a maximum speed of 1,320 miles per hour at altitude and may be capable of a combat radius of 2,485 miles in a hi-lo-hi mission profile.[11]  Normally the Backfire is armed with air to surface missiles.  Backfire is usually furnished with two AS-6 Kingfish rockets possessing an active radar guidance system and credited with a range of at least 135 miles.[12]

Backfire bomber units are deployed by the Soviet Union to threaten China, Japan, western Europe, and the North Atlantic.  Significantly, Backfires flying from bases in the Soviet Union can attack any point in the British isles.  Moreover such attacks can be made from the western approaches, thus substantially complicating British defensive measures.  A maximum ferry range of 5,000 miles enable Backfire squadrons to be rapidly switched from one theatre to another and concentrated to obtain utmost impact.[13]  According to The International Institute for Strategic Studies 145 Backfire machines were operational in 1980.[14]

It must be noted that Soviet Long Range Aviation

includes in its medium bomber force some 318 old and obsolete Tupolev 16s and 125 more modern and effective Tupolev 22s.[15] Neither warplane however is effective at low altitude penetration. The advent of Backfire has therefore endowed the Long Range Aviation medium range bomber force with a new dimension of striking power and effectiveness.

The newly developed strike capability of Soviet airpower, particularly of the tactical branch, highlights the strategic significance - and vulnerability - of Great Britain. England is the crucial reinforcement and supply base for the European theatre. Control of critical air and sealanes across the North Atlantic hinges upon the protective shield of air and naval forces based in the British isles. The strategic importance of Great Britain in thus enormous.

Could the Royal Air Force and its allies win a second air Battle of Britain? The air defense system of Great Britain is undergoing a sweeping expansion and modernization.[16] The northern and western approaches are being covered against enemy penetration. New mobile radar systems are being brought into operation to detect low flying Soviet intruders. The new radar systems are mobile and thus less vulnerable to attack than the older fixed radar sites. Command centers and communications lines are being shielded. Anti-aircraft missile units defend five airbases and other key installations.[17] A more effective air to air missile, the Skyflash, has been brought into service.

However much remains to be done. Many fundamental measures are currently in train to buttress the air defense of the United Kingdom. Elderly and obsolete Shackleton early warning radar aircraft will begin to be replaced in 1982 by eleven airborne early warning versions of the Nimrod maritime patrol aircraft. Air to air refuelling capability, needed to increase the interception radius of fighters on defensive patrol, will be augmented through the addition of nine VC-10s to the present aerial tanker force. More airbases will be protected by antiaircraft rocket formations. Indeed, the seven airfields employed by the U.S. Air Force in England will be guarded by British Rapier missiles operated by English troops.[18]

Most important of all, during the second half of the 1980s the current defensive fighters of the RAF will be supplanted by a new and far more effective interceptor, the Panavia F.2. The F.2 is a substantially modified air defense version of the twin engine, "swingwing", two seat Tornado low level strike warplane. At least 165 Tornado F.2s are to be secured by the Royal Air Force.[19] At least eight squadrons, committed to air defense of the United Kingdom region, will be equipped with the new machines.[20]

The Tornado interceptor has been specifically tailored to meet the unique demands of the British and northern European air arena. The sheer sweep of the key areas around the British isles - the eastern Atlantic, the Norwegian and North Seas - is indeed considerable. And enemy warplanes must be intercepted as far from the British coast as possible, to deny the foe the opportunity to unleash long range air to surface missiles against targets in England. English defenders must be able to cope with enemy warplanes streaking in at very low alti-

tudes and very fast. In the event that English ground
based defensive installations are severely damaged or
put out of action, interceptors must be able to contine
to fight autonomously. Inasmuch as electronic countermeasures warfare is a Russian specialty, British interceptors must be equipped to fight in an environment where
the duel between electronic countermeasures and counter-
countermeasures will be intense and decisive. It will
be necessary to disperse defending fighters to secondary
airfields. Interceptors must thus possess an on-board
ground power unit and an excellent short field landing
and take-off capability.

Range is a paramount concern in the British air defense region. Normal combat radius of the Tornado F.2,
flying from British bases, extends as far as the northern
coast of Iceland, Bodo in northern Norway, and Helsinki,
Finland.[21] Range can be considerably augmented by refuelling in air. Such range capability endows the F.2
with ample loiter and interception time.

The two man crew consists of a pilot who also handles
the weapons and a navigator who is responsible as well
for radar detection and tracking of the enemy and controll-
ing the complex counter-countermeasures electronic equipment. A two man crew is essential to cope with patrols
over long distances, to fight effectively independent of
ground control, and to function in miserable weather -
a 200 foot ceiling and pouring rain.

An advanced and highly sophisticated suite of elec-
tronics and avionics is the leading feature of the Tornado intercepter.[22] The new radar can detect enemy aircraft as far away as 100 nautical miles. A low level

intruder flying as low as 250 feet can be discerned and destroyed at a range of 25 miles.[23] A track while scan capability doubles the ability of Tornado to score lethal hits on enemy warplanes. And the extremely powerful radar is highly resistant to countermeasures jamming. Other means of target detection and identification are also provided.

A key feature of Tornado avionics equipment is a Tactical Evaluation Display which presents a detailed picture of enemy warplanes, cooperating defensive fighters, and other major participants. From this electronic display the Tornado crew can plan their sequence of interception attack.

Armament includes four Skyflash medium range missiles. Skyflash has been designed to counter the low level raider and to be effective in the intense electronic countermeasures environment. "Snap-up" attacks against high flying intruders can be readily unleashed as well. Two Sidewinder rockets and a 27 mm. machine cannon are provided for close combat.

Tornado F.2 can operate quite effectively from fairly short and widely dispersed airstrips. Employing thrust reversers Tornado can land in 1,200 feet. With normal load it can take off in 2,500 feet.[24] Ground power requirements are furnished by an integral auxiliary engine. All that will be needed to send the aircraft into action again will be fuel, ammunition, and oxygen.

The Tornado F.2 has been designed and configured to meet the demands of the interception role. It lacks the spectacular climb and maneuverability performance of such air superiority dogfighting machines as the McDonnell

Douglas F-15 Eagle or the General Dynamics F-16 Fighting Falcon. It should be noted, however, that Tornado is as fast as the MiG-23MF and can best the Soviet fighter in a sustained turn. And the handling qualities of the F.2 elicit universal pilot acclaim.

But the Tornado interceptor will not be in squadron service until at least the mid-1980's. Initial deliveries of the new fighter are not scheduled until 1984. In the crucial interim years the Royal Air Force must cope with circumstances that leave much to be desired.

Air defense of the British isles and surrounding waters currently is the responsibility of two squadrons of 24 Lightning fighters and five squadrons of 60 Phantom interceptors.[25] In the event of conflict, 9 Lightnings and 24 Phantoms and their instructor pilots from operational conversion units will be committed to the fray.

Operational since 1960, the Lightning is a very fast machine and features an exceptional rate of climb. Designed as a point defense, high altitude interceptor, however, it suffers from short range and carries only two air to air missiles. A very demanding machine to fly, the Lightning also requires many hours of complex maintenance.

The McDonnell Douglas Phantom was originally designed as a long range interceptor. A twin engine warplane with a crew of two, the Phantom is the present backbone of British air defense. Largely equipped with British electronic systems, the English version of the Phantom is also powered by Rolls Royce turbofan engines. The Phantom does possess the second crew member, heavy missile payload, powerful radar, and long range required for the interception mission. But the new generation of Soviet warplanes is the equal of the Phantom and possibly superior to it.

Far too heavy a burden of air defense falls upon too few fighters in the British isles. The English are quite aware of this crucial shortcoming. In an attempt to buttress British air defense some 90 Hawk advanced trainers will be armed with Sidewinder air to air missiles and employed in the low level point defense mission in time of emergency or conflict.[26]

Will this be sufficient? Government officials, testifying before the Defense Committee of the House of Commons in April of 1980 conceded that some 50 additional Royal Air Force interceptors, beyond present strength, are needed.[27] The sheer cost of securing first line aircraft and regular pilots, however, proved to be daunting. In their Second Report the Defense Committee of the House of Commons concluded that: "... the country is not adequately defended against a major attack..."[28]

What could possibly be done to improve the air defense of Great Britain in the years immediately ahead? To be sure, Lightning is too demanding of piloting skills and costly maintenance to be pressed into service in further numbers. Nor can Phantom strength be augmented. Phantom production lines in the United States are now closed. Moreover, from the maintenance standpoint, the largely British electronic systems and Rolls Royce engines render it impractical to transfer Phantoms from American air arms to the Royal Air Force.

What is to be done? Perhaps it would be possible to bring into service three squadrons of the older Hawker Hunter fighters manned and supported by reserve pilots and ground crew. In low level, last ditch defense of air bases, seaports, and other vital installations, the Hunter could still be a formidable adversary.

The Hawker Hunter is a relatively easy warplane to fly and has earned high marks for its fine handling qualities. Simplicity of maintenance is another Hunter feature vital to support by part-time ground crew. The Hawker fighter is armed with a quartet of 30 mm. machine cannon which can unleash a devastating hail of fire in close combat. And Sidewinder missiles can easily be added.

Moreover, a large number of Hunters in good operational condition are readily at hand. Almost 80 Hunters have been recently employed by tactical weapons training formations.[29] These machines are now being supplanted by the new Hawk trainers. More than enough Hunters should thus be available to equip three interceptor squadrons on a regular basis and provide ample spares.

There should be no insurmountable difficulty in finding enough former pilots with recent experience in the cockpit willing to serve as "weekend warriors" and enjoy flying a fighter which has a well deserved reputation as a pilots airplane. A much more difficult obstacle may be found in the attitude of the Royal Air Force toward "part-time" pilots in peacetime squadrons. In testimony before the Defense Committee of the House of Commons, government officials indicated that:

> "...the RAF has been firmly of the opinion that there was no need for reserves for flying tasks....To keep reservists up to date on modern aircraft is simply not worth doing..."30

To be sure, to put a part-time pilot in the cockpit of a Lightning fighter would be an invitation to swift disaster. Much depends on the aircraft in question. It should be emphasized that employment of Hunter formations would be an interim measure until sufficient regular Tornado units come into service.

Can American air arms contribute to the air defense of Great Britain? Assuredly England is a vital base for American tactical airpower. Long range strike F-111s, "tank-busting" A-10s, Phantoms and aerial tankers are among the American aircraft stationed at airfields in Britain. Most of these machines are configured for interdiction, close support, and reconnaissance missions. In time of dire emergency, however, some Phantoms would be committed to the fray in the interceptor role. But the bulk of American warplanes are dedicated to the crucial struggle over central Europe.

Possibly the most valuable contribution the United States could render would be the assignment of a second aerial tanker squadron to a permanent base in England. Such tankers are essential to enhance the range and operating flexibility of defensive interceptors. Enemy warplanes must be intercepted as far from the English coast as possible, before they can unleash air to surface missiles against targets in Britain.

Until the airborne warning version of the Nimrod is brought into service with the Royal Air Force, the United

States could commit several Boeing E-3A Sentry Airborne Warning and Control Systems aircraft to the defense of England. Aircraft can be detected by the E-3A some 250 miles away. The Sentry can track and identify 240 targets simultaneously.

Two squadrons of U.S. Navy Grumman F-14 fighters should be based in eastern Scotland. A twin engine machine with a crew of two, the F-14 Tomcat was conceived as a long range interceptor, designed to destroy fast Soviet bombers, and air to surface and surface to surface cruise missiles over water. Backfire bombers, attempting to attack England by the "back door" could be intercepted by F-14s far out over the Norwegian Sea. Stationed in Scotland, Tomcats can also intercept Soviet Naval Backfires on their way to raid convoys in the North Atlantic.

The F-14 formations would be supported by a squadron (four machines) of Grumman E-2C airborne early warning and electronic counter-countermeasures airplanes. The F-14 and E-2C constitute an integral interception team.

In an interception configuration range of the F-14 is some 2,000 miles.[31] It is also a swift machine, flying faster than the Phantom and Tornado and a bit faster than Lightning.[32] But the keys to the interception capabilities of the F-14 are the weapons control radar system and the long range Phoenix missile. Bombers can be detected at some 170 nautical miles.[33] Small cruise rockets can be picked up at approximately 65 nautical miles. The F-14 can track as many as 24 targets simultaneously and unleash Phoenix missiles against six of them. Maximum launch range of Phoenix is usually 52 nautical miles, although the missile can notch a "kill" at 110 nautical miles.[34]

It must be emphasized, however, that in a heavy electronic jamming situation - such as will be the case in defense of the British isles - the F-14 will need accompanying counter-countermeasures aircraft to be a viable weapon.[35] It must also be noted that the weapons control radar of the F-14 is optimized for operations over water, not land.[36] Hence F-14s would be committed to missions over the Norwegian Sea and the North Atlantic.

If these measures can be implemented, then the air defense of the British isles and surrounding seas can be assured in the dangerous years immediately before us.

[1] The International Institute for Strategic Studies, Strategic Survey 1978 (London: The International Institute for Strategic Studies, 1979), p. 33.

[2] "The Price of Power," Time (October 29, 1979), p. 28.

[3] "Eastern Bloc Augments Attack Force," Aviation Week and Space Technology (February 6, 1978), p. 57.

[4] Robert P. Berman, Soviet Air Power in Transition (Washington, D.C.: The Brookings Institution, 1978), p. 69.

[5] Clarence A. Robinson, Jr., "Soviets to Field 3 New Fighters in Aviation Modernization Drive," Aviation Week and Space Technology (March 26, 1979), p. 14.

[6] "Mikoyan Flogger," Air International (August, 1980), pp. 70-75 and 86-87.

[7] The International Institute for Strategic Studies, The Military Balance 1980 - 1981 (London: The International Institute for Strategic Studies, 1980), p. 12.

[8] Derek Wood, "The New Air Threat-Britain prepares," International Defense Review (October, 1977), p. 858.

[9] Bill Gunston, Consultant Editor, The Encyclopedia of World Airpower (New York: Crescent Books, 1980), pp. 334-335.

[10] The International Institute for Strategic Studies, The Military Balance.

[11] "Backfire," Air International (June, 1979), pp. 289-291 and 308.

[12] Gunston, op. cit., pp. 377-378.

[13] Ibid., p. 341.

[14] The International Institute for Strategic Studies, The Military Balance, pp. 10-11.

[15] Ibid., p. 10.

[16] Derek Wood, "The New Air Threat-Britain prepares," International Defense Review (October, 1977), pp. 855-859. and David A. Brown, "Britain Plan Expansion of Air Defense," Aviation Week and Space Technology (April 7, 1980), p. 48.

[17] Gunston, op. cit., p. 12.

[18] David A. Brown, "British Affirm Decision To Buy Trident SLBM's", Aviation Week and Space Technology (July 21, 1980), p. 25.

[19] Charles Gilson, "The Tornado F.2 ADV-UK's new interceptor," International Defense Review (No. 6, 1978), p. 865.

[20] Wood, op. cit., p. 859.

[21] Gilson, op. cit., p. 869.

[22] Ibid., pp. 866-868.

[23] Wood, op. cit.

[24] Gilson, op. cit.

[25] The International Institute for Strategic Studies, The Military Balance, p. 23.

[26] House of Commons, Minutes Of Evidence Taken Before The Defense Committee: Air Defense Of The United Kingdom (London: Tuesday 15 April 1980), p. 13.

[27] Ibid., p. 9.

[28] House of Commons, Second Report from the Defense Committee, Session 1979-80: Statement On The Defense Estimates 1980 (London: 23 April 1980), p. 5.

[29] Gunston, op. cit.

[30] House of Commons, Minutes Of Evidence Taken Before The Defense Committee: Air Defense Of The United Kingdom (London: Tuesday 15 April 1980), p. 15.

[31]Gunston, op. cit., p. 194.

[32]Ibid.

[33]Walter Maguire, "The Phoenix Factor," Air Combat (March, 1979), p. 62.

[34]Clarence A. Robinson, Jr., "F-14 Demonstrates Agile Aerial Combat Capabilities," Aviation Week and Space Technology (November 29, 1976), p. 52, and Maguire, ibid., p. 61.

[35]Clarence A. Robinson, Jr., "U.S. Reexamines Fighter Needs," Aviation Week and Space Technology (April 23, 1979), p. 18.

Chapter II

Challenge on the High Seas: The Soviet Navy

Among the key Soviet armed forces programs is the development of a major navy, technologically advanced and capable of operating in any far flung corner of the seas of the world. In the course of the 1970s the leadership of the Soviet navy has waged a remarkably successful struggle to secure a much more important position in the hierarchy of Russian armed forces and win extra funding for fleet construction. Consequently, the Soviet navy is now a formidable instrument indeed. The International Institute for Strategic Studies indicates 287 major surface combat vessels and 344 submarines (162 of them nuclear fuelled) in the array of active units in the Soviet navy in 1980.[1]

Many new features are to be seen in current Russian naval development. A massive strategic nuclear force is now based in submarines and plays a crucial role in Soviet nuclear strategy. Eighty seven Russian submarines carry 1,003 submarine launched ballistic missiles.[2] Nor has the surface fleet been neglected. Michael McGuire, an authoritative English analyst of Soviet naval affairs, comments incisively:

> The new surface ship programs represent an increase both in the number of ocean-going warships delivered each year and in the size of the various ship types. The end-product will be a much more powerful fleet, with a greatly enhanced general-purpose capability.[3]

And MccGuire concludes:

> The Navy's political standing has increased significantly over the last decade and may still be waxing. And naval design criteria have shifted from short-term survivability to sustaining combat operations for the duration of a war. This means that, for the first time, wartime requirements will generate a general purpose Navy with a true worldwide capability, suitable for use as an instrument of state policy in peacetime.[4]

The technological prowess of the Soviet navy is displayed in a variety of new warships. A unique and powerful 43,000 ton aircraft carrier equipped with vertical take-off and landing fighters and surface to surface missiles now carries the Soviet flag on the high seas. Two of these Kiev class carriers are in service and two more are under construction.[5] Kenneth McGruther underlines the significance of Kiev:

> Kiev is the first truly general-purpose ship of the Soviet Navy. It is capable of long-range strikes against surface targets with its SS-N-12 missiles; anti-submarine warfare; command and control; reconnaissance, close air support, or even local air superiority with its VTOL aircraft; and self-defense with its close-in weapons . . . Employed in conjunction with the newest of the other oceangoing classes of ships in the Soviet Navy, . . . Kiev provides the nucleus of a multimission, potent naval task group . . . With Kiev, a measure of sea control over areas of the Soviet Navy's choosing has now become feasible for the Soviet Navy.[6]

Advances in submarine technology are represented in the Alpha class of nuclear fuelled attack submersibles. Constructed of titanium, the Alpha is the swiftest submarine in the world at 40 knots and the deepest diving, operating at more than 600 meters depth.[7] "Bearing in mind the Soviet's capacity for innovation and their penchant for adopting unconventional means to outflank a superior capability," Michael MccGuire emphasizes,

> We should expect Alpha to be only the first of a series of advances, which could challenge our technological lead in the submarine field, and may also affect our future anti-submarine capabilities.[8]

A spectacular development in Soviet surface warships is to be seen in the 25,000 ton battle cruiser Kirov which joined the fleet in 1980. Nuclear fuelled propulsion machinery confers upon the Kirov high speed and very long range. In contrast to former Soviet warships, the Kirov possesses large magazines furnishing ample reload for her weapons and hence capability for sustained combat.[9] And nuclear fuelled aircraft carriers, perhaps as large as 70,000 tons, are under construction.[10]

A new appreciation of the value of sea power on the part of Russian leadership is reflected in the sharply rising prominence of the Soviet fleet. The Soviet regime increasingly understands the influence naval forces can have in peacetime. Fleets deployed in sensitive areas of the world can exert substantial political impact in support of Soviet interests. A case in point is to be seen in the commitment of 26 Soviet naval vessels to the Indian Ocean, astride the vital oil lifeline to the United States, western Europe, and Japan.

A continuing concern of Soviet leadership is the abrasive and ongoing conflict with Communist China. Russian armed forces in the Far East are heavily dependent upon the Trans-Siberian railroad for supplies and reinforcement. That rail link lies perilously close to the Chinese border and could be readily severed by limited enemy attacks. In the event of conflict with China, the

Russians might well need to rely upon their rapidly growing fleet of freighters and tankers to transport needed supplies to Soviet forces locked in battle in the Far Eastern theatre. Such seaborne transport would be compelled, however, to run the gauntlet of the Communist Chinese attack submarine force, third largest in the world. Hence the need for a Soviet fleet with global reach.

Above all other considerations, however, is a fundamental shift in Soviet military strategy and doctrine. From preoccupation with exclusively nuclear warfare Soviet thinking has broadened to encompass in future conflict a conventional phase and the possibility of a protracted war. As Michael MccGuire points out:

> Surface ships now had to be capable of the sustained operations needed to gain and maintain command of a large sea area such as the Norwegian Sea, and this required long endurance, large magazine loads and an underway replenishment capability.[11]

In addition to these peacetime and wartime tasks, the Soviet navy is entrusted with a number of other vital missions. A considerable proportion of Soviet strategic nuclear striking power is going to sea in new and much more potent and long ranging missiles based in huge nuclear fuelled submarines. This is particularly in evidence with the Delta class SSBN. The Deltas are armed with the SS-N-8 missile, a strategic rocket whose range is 4,300 nautical miles. Such an increasingly vital force must necessarily be shielded against any American attempt to eliminate it in the event of war. Such protection is especially crucial, Kenneth McGruther points out, because:

> Most commentators have come to the conclusion that over the course of the last decade the role of the Soviet Navy's strategic force has been fundamentally changed to one of carrying out 'deterrence' in war, conducting intrawar bargaining, and influencing the peace talks at the end of the war.[12]

Such a strategy of withholding strategic nuclear strength in reserve underlines the increased significance of the Soviet navy, but it also imposes a heavy responsibility upon that fleet. Sanctuary areas must be found wherein Soviet missile submarines can be effectively protected and all enemy efforts to destroy them can be thwarted. Such "bastions," as they are denoted by the Russians, are to be found in the Sea of Okhotsk, the Barents and Greenland Seas, and the Norwegian Sea. Thus the Soviet navy is expected to dominate these waters out to a "line" stretching from Greenland south to Iceland and thence to the northern tip of the British isles, the "GIUK Gap." Ironically, however, if the Soviet fleet succeeds in this crucial _defensive_ mission it will achieve a superb _offensive_ position as well. For the Soviet navy would then be poised directly above the sealanes of the North Atlantic. Moreover, control of the Norwegian Sea entails the Soviet seizure of such key islands as Svalbard and of the northern reaches of Norway. Norway is of course a member of NATO. And the United States, as a member of that western alliance, is committed to defend the national and territorial integrity of its allies. Ironically the mandates imposed by developing strategic forces entangle the two superpowers in a web of conflict.

How can the Soviet fleet attempt to protect its

strategic missile submarines against enemy attack? The Russians must contend against American and allied long range anti-submarine patrol planes, carrier based anti-submarine aircraft, helicopters flying from destroyer decks, and nuclear fuelled attack submarines. Under these circumstances it is not surprising that the bulk of the high seas Soviet surface fleet, much of the Soviet Naval Air Service, and the most advanced attack submarines are dedicated to frustrating the efforts of these NATO forces. This is, of course, the situation at present. As the Soviet navy brings into service an increasing number of supply vessels and much more powerful warships it may, in the future, carry out its defensive role and world wide naval operations simultaneously. Currently the Russians can pose lethal threats to western patrol planes and aircraft carriers. Coping with American and British nuclear fuelled attack submarines, however, is very difficult indeed. But western attack submarines must enter the Soviet sanctuaries to seek out Russian strategic missile submarines and thus encounter Soviet underwater sensors and defensive systems. The most advanced Soviet nuclear fuelled attack submarines, such as Alpha, can then be brought into play in the underwater duel. The Soviet effort is summarized by John Herzog:

> Placing the SSBNs inside the sanctuaries and by using new and sophisticated surface ASW ships, attack submarines, and BADGER and BACKFIRE aircraft to support them, the Soviet leaders could ensure the survivability of their strategic threat to the US.[13]

What then of Soviet operations against the North Atlantic sealanes? According to Fleet Admiral Sergei G.

Gorshkov, Commander-in-Chief of the Soviet Navy, writing in a recent edition of the Soviet Military Encyclopedia, operations against enemy sealanes are outranked in importance only by the Navy's strategic nuclear submarines and the forces needed to protect them.[14] And Russian words are backed up by action. Soviet naval maneuvers through the 1970s displayed sharply increased attention to air and submarine operations against enemy convoys. Donald Daniel emphasizes:

> For many commentators some of the most significant and surprising aspects of OKEAN 75 were activities which suggested heightened Soviet interest in sea lines of communications (SLOC) interdiction.[15]

Far more compelling, however, than doctrine or exercises are the new developments in the nature of modern conflict. The new emphasis by the Soviet Union upon conventional warfare and protracted conflict strongly enhances the role of disruption of enemy sealanes. This is particularly true because massive American troop reinforcements and war materials play such a crucial part in any conventional conflict in the European theatre. And a Soviet invasion of Norway and Denmark, an integral part of the new Soviet strategic picture, would swiftly bring British, Canadian, and American forces into the Norwegian Sea on the way to succour their Scandinavian allies.

The dependence of western European economies upon seaborne supply of vital raw materials is another fundamental factor increasing the value of the sealanes. Such western European vulnerability is an alluring invitation to Soviet attack. The Russians might indeed be tempted to seek a quick collapse of the western Europeans through a dramatic threat to their sea lifelines.

What forces can the Soviet navy employ against the North Atlantic sealanes? For the interdiction of sea lines of communication the Russians rely upon a team of submarines and long range fast warplanes armed with air to surface missiles. As John Herzog points out: "These units are well-equipped to venture out into the Atlantic and Pacific to search out, with the help of reconnaissance satellites, Western task forces and convoys."[16]

In 1980 the Soviet naval array included 135 ocean-going diesel-electric powered and 46 nuclear fuelled attack submarines.[17] In addition, there are 16 diesel-electric engined and 45 nuclear fuelled submarines armed with cruise missiles.[18] The missile in service, the SSN-7, can be fired submerged and ranges out to 30 miles.[19] To be sure, many of these submarines would be pitted against American carrier task forces and other surface warships. However Soviet cruise missile equipped submarine employment against convoys and merchant vessels cannot be dismissed.

Backbone of the diesel-electric attack fleet is the Foxtrot class submarine; sixty of which are currently in service; plus ten of the similar but larger Tango class.[20] Foxtrot is a well-proven and strongly built conventional submarine. Operational range of Foxtrot is 11,000 nautical miles and it is armed with ten torpedo tubes.[21]

A particularly menacing predator upon the sealanes is to be found in the nuclear fuelled Victor class. Featuring a "teardrop" hull form, the Victor displaces 5,700 tons submerged. Siegfried Breyer and Norman Polmar claim

a maximum speed for Victor II of 33 knots underwater.[22]
The Victor is armed with light torpedo tubes and can also
unleash tactical surface to surface missiles. Twenty
six Victors are presently operational and more advanced
versions are under construction. To be sure, a number
of the fast Victors are assigned to protecting SSBNs in
the northern sanctuary seas or attempting to trail American SSBNs. Nonetheless, even a proportion of the Victor
force loosed upon convoys could be devastating.

It should be noted that most Soviet submarines feature a special hull coating which absorbs some sonar
echoes, and renders sonar detection more difficult.[24]
But the Russians have not chosen to invest in sound-deadening equipment and materials to make their submarines quiet. Indeed, Soviet submarines are notoriously
noisy. Walter Pincus indicates that American underwater
listening systems ". . . can distinguish the peculiar
sounds put out by each particular Soviet sub - much like
the voice of an individual."[25]

However, the Soviet strategy of submarine operations
differs markedly from the pattern universally employed
in the west. Western submarines normally operate alone,
employing their own sensor equipment and patrolling wide
ocean areas in search of their quarry. In contrast,
Soviet submarines are deployed to pre-determined stations
astride the logical routes taken by western task forces
and convoys. As William J. Ruhe points out:

> Lying somewhere in the oceans, the Soviet submarine
> would be minimizing chances of being detected by
> means of very low-speed operations or through the
> use of auxiliary means of propulsion . . . In fact,
> the Soviet submarine would be in the role of artil-

lery, directed swiftly into action by a remote commander who has gained his information on enemy targets from other sources.[26]

What are these "other sources"? They are the wide array of Soviet military space systems and sophisticated communications networks. Particularly useful are the electronic ferret and radar reconnaissance satellites. And military navigation satellites furnish missile launching solutions and mid-course guidance. The close integration of military systems in space and Soviet warships is underlined by John Herzog:

> To take advantage of the scope of electronic warfare, they have developed systems whereby the surface fleets of the west can be reconnoitered continuously . . . With reconnaissance satellites in the usual 90 minute orbit, the coordinates of surface targets can be determined and the information passed on to CHARLIE or VICTOR submarines for targetting purposes. The same information would be passed to Moscow Naval Headquarters where the decision would be made on positioning the submarines and from where the firing order would come.[27]

The other partner in the Soviet anti-shipping strike team is the medium bomber in naval air service. Spectacular developments in Russian surface warships and submarines have attracted much attention. The steady evolution of the AV-MF - Soviet Naval Aviation - has aroused less interest. Yet Paul Murphy points out that:

> The past decade and a half clearly represents the most important period of development for the Soviet Navy's aviation component. . . . Under the supervision of Admiral Gorshkov and the direction of General-Colonel I. Borzov, then the Chief of Naval Aviation, the Navy's traditionally subordinate air arm underwent a rejuvenation in the mid-1960s and

by 1969-70 it emerged as a prominent component of the Soviet Navy.[28]

In 1980 some 775 aircraft and helicopters were in service with the naval air arm.[29] A wide variety of warplanes are to be found in this air force including anti-submarine patrol planes, long range reconnaissance machines, and carrier based vertical take off and landing fighter bombers.

In the anti-shipping strike role the medium bomber represents the key Soviet naval air weapon. For this purpose 250 older and slow Tupolev-16s remain in service with AV-MF.[30] Badger, as it is code named in the west, has a radius of 2,000 miles and is armed with two air to surface missiles. A more modern machine, the Tupolev-22 or Blinder features a maximum speed of 1,000 miles per hour and a radius, with 250 miles of supersonic dash, of 1,750 miles.[31] Approximately 45 Blinders are in naval service.[32] Ranges can be substantially extended by in-flight refuelling. No less than 74 Badgers serve as tankers.

Much more menacing however is the modern Tupolev Backfire - 70 of which comprise the most deadly strike bomber force in Soviet naval aviation.[33] A fast machine Backfire is capable of 1,320 miles per hour at altitude and 650 miles per hour at sea level.[34] Backfire features an estimated unrefuelled radius, with 200 miles of supersonic dash and 200 miles at low altitude, of 2,485 miles.[35] Thus Backfire can be unleashed upon shipping lanes from Newfoundland to the Azores. Refuelled in air, Backfire can range from northern New England to the west coast of Africa.

Primary armament for Backfire is the AS-6 Kingfish air to surface missile. Kingfish carries a 2,205 lb. warhead a distance of 135 miles in a low altitude mission profile. If most of the distance to a target can be covered at high altitude, then range can be as far as 380 nautical miles. The rocket does, of course, require mid-course guidance by another aircraft or surface vessel. As the rocket closes on its target a maximum speed of Mach 3.5 may be reached.[36] Active radar guidance then homes the missile onto its quarry.

Little wonder that a former Royal Air Force Air Staff Chief, Andrew Humphrey exclaimed: "Fast, wide-ranging aircraft like Backfire, armed with stand-off missiles, may soon become an even greater danger to allied shipping than the relatively slow-moving Russian submarines!"[37] The point is reinforced by Peter Rasmussen who comments on Badger and Backfire: ". . . their range and the relatively short distance from the Kola Peninsula will allow them to conduct immediate strikes against the sealanes, provided that reconnaissance and target acquisition are secured."[38] Such needed target information is furnished by long range patrol planes. The naval air arm operates 115 aircraft in the maritime reconnaissance role.

It should be noted that Badger, Blinder, and Backfire squadrons can be swiftly assembled from all corners of the Soviet Union and concentrated on one major theatre. As Peter Rasmussen indicates:

> ". . . in a conflict limited to the Northern Atlantic, SNAF formations under the command of the two north western fleets could receive reinforcements from the SNAF in the Black Sea or even from Pacific Fleet units."[39]

The Russian navy may be entering upon the most formidable phase in its long history. Clearly, Soviet submarines and warplanes constitute a serious threat to the sealanes of the North Atlantic.

[1] The International Institute for Strategic Studies, *The Military Balance 1980-1981* (London: The International Institute for Strategic Studies, 1980), p. 9, 11.

[2] *Ibid.*, p. 9.

[3] Michael MccGuire, "A New Trend in Soviet Naval Developments," *International Defense Review* (No. 5, 1980), p. 677.

[4] *Ibid.*, p. 680.

[5] The International Institute for Strategic Studies, *op. cit.*, p. 11.

[6] Kenneth R. McGruther, *The Evolving Soviet Navy* (Newport, Rhode Island: Naval War College Press, 1978), p. 78.

[7] Gowri S. Sundaram, "ASW - the key to sea control," *International Defense Review* (No. 3, 1980), p. 367.

[8] MccGuire, *op. cit.*, p. 679.

[9] "Soviet Nuclear-Powered Surface Warship Cruises Baltic," *Aviation Week and Space Technology* (October 20, 1980), p. 31.

[10] MccGuire, *op. cit.*

[11] *Ibid.*, p. 676.

[12] McGruther, *op. cit.*, pp. 57-58.

[13] John Herzog, "Perspectives on Soviet-Naval Development: A Navy to Match National Purposes," in Paul J. Murphy, ed., *Naval Power in Soviet Policy*, *Studies In Communist Affairs*, *Volume 2*, published under the auspices of the United States Air Force (Washington, D.C.: Government Printing Office, 1978), p. 53.

[14] Rear Admiral Sayre A. Swarztrauber, "The Potential Battle of the Atlantic," in Frank Uhlig, ed., *Naval Review 1979* (Annapolis, Maryland: U.S. Naval Institute, 1979), p. 114.

[15] Donald C. Daniel, "Trends and Patterns in Major Soviet Naval Exercises," in Paul J. Murphy, ed., op. cit., p. 227.

[16] Herzog, op. cit., p. 55.

[17] The International Institute for Strategic Studies, op. cit.

[18] Ibid.

[19] Jean Labayle-Couhat, ed., Combat Fleets Of The World 1980/81: Their Ships, Aircraft, and Armament (Annapolis, Maryland: The United States Naval Institute, 1980), p. 527.

[20] The International Institute for Strategic Studies, op. cit., p. 11.

[21] Labayle-Couhat, op. cit., pp. 548-550.

[22] Siegfried Breyer and Norman Polmar, Guide to the Soviet Navy, Second Edition (Annapolis, Maryland: Naval Institute Press, 1977), p. 160.

[23] The International Institute for Strategic Studies, op. cit.

[24] Labayle-Couhat, op. cit., p. 539.

[25] Walter Pincus, "Hill Study Cites Vulnerability of Soviet Missile Submarines," The Washington Post (January 10, 1979), p. A5.

[26] William J. Ruhe, "Missiles Make ASW a New Game," Proceedings: United States Naval Institute (March, 1980), p. 73.

[27] Herzog, op. cit., p. 54.

[28] Paul J. Murphy, "'Morskaya Aviatsiya' (Soviet Naval Aviation): Its Development, Capabilities, and Limitations," in Murphy, op. cit., p. 180.

[29] The International Institute for Strategic Studies, op. cit.

[30] Ibid.

[31] Bill Gunston, Consultant Editor, The Encyclopedia of World Airpower (New York: Crescent Books, 1980), p. 339.

[32] The International Institute for Strategic Studies, op. cit.

[33] Ibid.

[34] Gunston, op. cit., p. 341.

[35] "Backfire," Air International (June, 1979), p. 308.

[36] Gunston, op. cit., pp. 377-378.

[37] "Backfire," op. cit.

[38] Peter H. Rasmussen, "The Soviet Naval Air Force: Development, Organization and Capabilities," International Defense Review (No. 5, 1978), p. 694.

[39] Ibid.

## CHAPTER III
## Control of the Air at Sea

How can the western alliance cope with the twin challenges of Soviet naval airpower and submarines to the sealanes of the North Atlantic? The most immediate, potentially devastating and rapidly growing threat is posed by the Backfire missile-firing bomber. What can be done to thwart the depredations of Backfire?

Can fighter units based on shore ensure protection for task forces and convoys far out at sea? Are land bases a less costly and more efficient means of securing effective tactical airpower? The limitations of land based fighter cover are incisively underlined by M. Forrest:

> To maintain a Combat Air Patrol of two Phantoms over a force 450 miles from an airfield would need six serviceable aircraft and 12 crews . . . . Move the force out to 700 miles or so and a squadron of 14 aircraft at 70 per cent serviceability is required.[1]

And the vital part played by fighters against enemy warplanes is concisely portrayed:

> Fighter aircraft have an important job to do even before any attack is launched . . . . The shadowing aircraft (enemy), keeping outside SAM range, is there to provide continuous briefing for the raid leader and it is the fighter's task to shoot down the shadower before the raid comes in. Secondly, if there are no fighters, the bombers can improve their chances of success by coming in close enough (a range bracket of 70-100 miles, depending on missile radar frequency) to allow the missile head to lock on to the target before launch . . . . When the raid comes, only fighters can pre-empt attack by shooting down the bombers before missile release . . .[2]

Thus there can be no substitute for carrier based fighter aircraft.

Moreover carrier based warplanes are needed to protect western interests in far corners of the world. Many of the natural resources upon which western economies depend flow from the Persian Gulf area and southern Africa. In these regions airfields are few and quite inadequate to sustain operations by sophisticated aircraft. Politically, these areas are explosively volatile and unstable. Under such circumstances, carriers are the swiftest and most reliable way to bring airpower to bear in the crises that convulse these vital areas.

Studies have demonstrated that the giant attack carrier is the most powerful and efficient instrument of airpower at sea.[3] The large carrier benefits from sheer economy of scale, and can carry many warplanes and much fuel and munitions. Such warships also feature extensive repair and maintenance shops. And the super carrier is very fast and long ranging. Although no warship can be invulnerable, the giant carrier can absorb tremendous punishment and is extremely difficult to put out of action.

The large attack carrier is of course quite costly. Much of the striking power of the fleet is concentrated in a relatively small number of key warships. And even the huge carrier is vulnerable to nuclear weapons. Hence the loss of a super carrier would be a serious blow.

In 1980 twelve first line attack carriers were to be found in the American fleet.[4] One super carrier is undergoing major overhaul and another has just begun comprehensive modernization. One nuclear fuelled attack carrier of the Nimitz class is under construction and is expected to enter service in 1982. And funds have been appropriated for an improved Nimitz class carrier. A series of thorough modernization overhauls, the Service Life Extension Program, will ensure the viability of the super carrier fleet through the 1990s.

American carrier based naval and Marine fighter strength in 1980 numbered 168 Grumman F-14As and 288 of the much older McDonnell Douglas F-4 Phantoms.[5] The cutting edge of carrier based airpower was to be found in 300 Vought A-7 Corsair II clear weather attack planes, 170 Grumman A-6 all weather attack machines, 80 McDonnell Douglas A-4M light attack aircraft, and 78 British AV-8A V/STOL fighter bombers.

The aging Phantom fighters and the Corsair attack machines (limited to clear weather operations) are scheduled to be supplanted by a new fighter/attack plane, the McDonnell Douglas F-18 or Hornet. However the Hornet is only now undergoing testing. And the expected cost of the Hornet program has skyrocketed to an estimated 30 billion dollars.[6] It will thus be the most expensive weapon program in the navy.

What capabilities does American naval airpower really need? How can these capabilities be realized as swiftly as possible?

The attack carrier must be protected against attack by fast bomber and missile. This essential task can be most effectively performed by the F-14. Substantially more of these advanced "swing-wing" machines should be secured. Indeed the Tomcat ought to be the navy's fighter.

For offensive punch, the navy needs a warplane able to operate in the worst weather conditions, function in a massive electronic situation, and possessing long range. Fortunately, such a machine exists in the Grumman A-6E Intruder.[7] Electronically a very sophisticated aircraft, the Intruder is equipped with an all weather navigation system, and an extremely accurate computer attack system. And the A-6E can carry a substantial payload over a radius of some 950 miles. Considerably more Intruders are needed (shipboard squadrons are currently under strength) and should be procured. Intruder should constitute the prime offensive weapon of naval airpower.

Inasmuch as F-14 and A-6 can quite effectively fulfill the roles envisaged for the F-18 that costly program could be cancelled and the funds employed to purchase Tomcats, Intruders, and - an improved model of the V/STOL fighter bomber. A warplane is needed to provide close support to the Marines and other ground troops.

The British Aerospace Harrier V/STOL fighter bomber has been in service with the Royal Air Force (151) and the United States Marine Corps (110) for some ten years and is a thoroughly tested and proven aircraft. It is also operational with the Spanish Navy, and a special version, the Sea Harrier, is entering service with the Royal Navy.

Quite different from conventional aircraft, the Harrier enjoys many advantages. It can fly from small vessels with short decks - no costly and cumbersome steam catapults and arrester gear are necessary. On shore it can operate from small and primitive airstrips and other rudimentary sites. The Harrier is easy to maintain, service, and rearm. And it can operate with a ceiling as low as 200 feet and visibility of ½ mile.

To augment range and payload the Harrier is most frequently flown with a short take-off run and landed in the vertical mode. In Marine Corps service the AV-8A can carry a four ton payload of fuel and munitions over a radius of 225 nautical miles, can fly more than six missions a day, and can be on its way within 1½ minutes of a request.[8]

Harrier is thus a particularly useful fighter bomber in the close support mission. And it is a formidable opponent in air to air combat. A very high ratio of engine power to weight endows the Harrier with excellent acceleration and a superlative rate of climb. Exceptional maneuverability is conferred by the ability to rotate the thrust nozzles while in level flight.[9]

In the North Atlantic arena the main opponents of Harrier are Soviet Badger, Blinder, and Backfire bombers and reconnaissance machines. Harrier easily exceeds the top speed of the Badger. Blinder and Backfire are fast machines - but when laden with externally mounted missiles their speed is constricted to a normal cruise of 560 miles per hour, well below the maximum Harrier speed of 737 miles per hour. Backfire - clean - can outrun Harrier at altitude, but at low level tops out at 650 miles per hour.

The Sea Harrier in Royal Navy service can fly an interception mission out to a 400 nautical mile radius.[10] In the reconnaissance role, Sea Harrier can scout at low level 20,000 square miles of ocean in one hour. And Martel or Harpoon missiles can be unleashed against enemy warships. Thus Harrier is a versatile warplane, capable of fulfilling enough roles to make a "Harrier carrier" a viable concept.

Critics of V/STOL aircraft have argued that such machines are inferior in range and payload to conventional planes. To be sure, the Harrier is not at all suited for long range penetration or offensive, high speed fighter operations. However, the installation of an upward curve in the front end of a flight deck can give extra impetus to a short take-off. Enjoying the advantage of such a "ski-jump" take-off, the Harrier can carry an extra ton of payload or gain a 50 percent increase in radius of action.[11]

The most incisive assessment of Harrier, however, is voiced by Roy Braybrook who outlines the dilemmas levied by V/STOL operation upon engine design:

> In order to achieve a high thrust/weight ratio engine, the designers of the Pegasus chose a moderately high bypass ratio (in comparison with the normal order of bypass ratio for high performance aircraft), which inevitably involves a rapid thrust decay with forward speed. As a result, the maximum attainable speed of the Harrier is restricted to something in the region of 600 knots, a speed within the reach of conventional aircraft of much lower static thrust/weight ratio.[12]

Braybrook also emphasizes the necessity for a more powerful engine. As more equipment and weight has been added to Harrier the margin of power available for operations in hot weather conditions and high airfields has waned considerably.[13] A particular handicap for the Harrier in the fighter role is the difficulty in a vertical landing with the weight of unexpended missiles, especially in hot and high situations.

Even so, flexibility, versatility, and ability to operate from small and austere surfaces make the V/STOL warplane an attractive asset. And the employment of V/STOL aircraft could secure more airpower at sea on a much wider variety and larger number of vessels. Such a dispersal of sea based airpower could partially offset the present concentration of naval aviation in a dozen super carriers. The catastrophic impact of the loss of an attack carrier would be lessened by the survival of aircraft in dispersed and smaller aviation ships.[14]

Above all, airpower will be more widely available at sea for such purposes as amphibious operations and sealane security. It will be desirable to build a fleet of smaller carriers to operate the V/STOL aircraft. But Harriers can be brought into operation immediately on a wide array of ships currently in service. Even old carriers in reserve can be taken out of "mothballs" and readily adapted, with a minimum of crew, to Harrier operations.

The Marine Corps would certainly welcome Harrier contingents aboard the Tarawa class amphibious assault ships. Indeed, the Guam has functioned as a carrier for Marine AV-8As. And the Okinawa, an Iwo Jima class amphibious assault helicopter carrier, has sustained twelve Harriers in operation. Five Tarawa class and seven Iwo Jima class vessels are presently in service.[15]

It might be possible to employ the Oriskany, now in reserve, as a V/STOL carrier. Weighing 40,600 tons at full load, the Oriskany is swift and equipped with fairly modern electronic systems. Some 40 Harriers and a contingent of anti-submarine Lockheed S-3A Viking planes and Sikorsky SH-3G Sea King helicopters could be carried. Paul S. Trible, member of the House of Representatives Armed Services Committee, has pointed out: "For $185 to $200-million, this ship could be overhauled and have a service life of 15 years."[16]

What new V/STOL fighters could be brought into production? Now beginning to enter service with the British fleet, the Sea Harrier is a navalized version of the machine operated by the Royal Air Force. Major modifications include a modified wing design, a new radar and navigational computer, and a raised cockpit for improved visibility.[17] The Sea Harrier is optimized for interception and reconnaissance missions. It will also be equipped with Sea Eagle air to surface missiles to be employed against warships.

McDonnell Douglas has developed and tested, under license, an improved version of the AV-8A to meet Marine Corps needs. The Marine Corps would like to procure 336 of these AV-8B advanced Harriers.

Weight is substantially reduced in the AV-8B through the employment of carbon fiber, more than 23 percent of the airframe being graphite epoxy. Carbon fiber composites are not only light in weight but are also long lasting and do not corrode.[18] A new supercritical airfoil wing and redesigned engine intakes reduce cruise drag and thus increase range. A variety of devices are built into the bottom of the fuselage to improve vertical lift. Among the features incorporated in the avionic suite are a laser and TV tracker, the mounting of many essential controls on the throttle and stick, and digital displays which highlight the instrument panel.

All these features result in a substantial increase in AV-8B performance. As Graham Warwick summarizes:

> AV-8B is essentially a bomb truck. Payload/range performance is at least double that of the AV-8A . . . . The AV-8B is designed to lift 28,750 lb- including 5,500 lb of fuel and 16 Mk82 bombs (9,120 lb)-from a 1,000 ft, sea-level runway on a tropical (90°F) day.[19]

The AV-8B is thus optimized for cruising at medium altitude in a high-low-high mission.

Very different requirements dictate the choice of a successor to the current Harrier in Royal Air Force service. The RAF is committed to operations in the densely packed and immensely lethal central front of continental Europe. Under such circumstances a tactical fighter bomber must be capable of high speed at low altitude - to have any chance of survival in the face of massed anti-aircraft weaponry. Maximum maneuverability is also deemed essential to cope with swarms of nimble Russian fighters.

To meet these demands a new wing design is undergoing wind tunnel testing for the proposed Harrier Mk.5.[20] Lift is also substantially increased through an extension fitted to the leading edge of the wing root. A singular advantage of this new wing design is that it can be retrofitted to Harriers presently in service.

None of these developments, however, address a fundamental problem - the need for a substantial increase in engine power. An AV-8B Plus has been suggested incorporating radar and a modified Pegasus engine providing 23,000 pounds of static thrust.[21] Whatever form a future

Harrier may take, a more potent engine is necessary.

What new carriers are designed to operate V/STOL aircraft? Although the Italian Navy does not intend to bring such aircraft into service, the firm of Italcantieri is building a helicopter carrier readily adaptable to V/STOL operations. Fully laden at 13,250 tons, the Garibaldi is fast and well designed.[10] The superstructure as well as the hull is built of steel. Special attention has been paid to stability for operations in severe weather. Heavily armed, the Garibaldi mounts anti-aircraft missiles and cannon, anti-submarine torpedo tubes, and surface to surface missiles. Thus it can operate independently, as well as part of a task force.

Equipped with a "ski-jump" ramp, the Garibaldi is on offer to the navies of Australia and Brazil as a V/STOL carrier. In that role it can hangar eight to ten Harriers plus one to three Sea King helicopters. Although tightly packed and cramped, the Garibaldi is the best armed and most efficient small V/STOL carrier.

Under construction for the Spanish Navy, which does operate a handful of AV-8As, is a 14,300 ton V/STOL carrier. This vessel is simple, relatively slow, and furnished only with short range gatling type anti-aircraft cannon armament.[23] It is to be protected by an escort of Perry class frigates, three of which are to be built in Spanish yards. Nineteen Harriers and Sea King helicopters will be the complement of this new carrier. This Spanish warship will essentially be an austere floating airbase.

Currently the only new V/STOL carrier in service is the Invincible of the Royal Navy. It will be joined in the early 1980s by the Illustrious and Ark Royal. The design of the Invincible class underwent many permutations, reflecting the fierce debates over naval roles and missions that rang through the corridors of Whitehall. Originally intended as a helicopter carrier for anti-submarine operations, Invincible proved readily adaptable to V/STOL operations.

At 19,810 tons fully laden, the Invincible is a fast and spacious warship.[24] Armament is limited to medium range anti-aircraft missiles. Five Sea Harriers and nine Sea King helicopters represent the present air component. Far more Harriers, however, could easily be accommodated by Invincible. A "ski-jump" ramp is fitted at the end of the flight deck.

Moreover, Invincible is a versatile warship. Designed to be the command center of a substantial task force, the Invincible features commodious command facilities and an elaborate and comprehensive communications system.[25] Sufficient space is also furnished to carry a battalion of Royal Marines, some 750 troops.

American needs, however, may best be met by adding more Tarawa class amphibious assault ships to the fleet. Weighing 39,300 tons fully loaded, these vessels are versatile and well designed.[26] The Tarawa is fully outfitted with command and communications equipment. Thirty assault helicopters and 1,900 Marines can be carried. And it is fairly well armed with anti-aircraft missiles and guns. Certainly a Tarawa can support a substantial number of Harriers and Sea King helicopters.

Four Tarawa class carriers, dedicated to V/STOL operations, should be ordered. Funds with which to purchase the Tarawas can be secured through a sharp reduction in the planned acquisition of cruisers mounting the Aegis anti-aircraft system. Tentatively pegged at 24 vessels, the Aegis program would cost almost 20 billion in fiscal year 1980 dollars.[27] Twelve Aegis cruisers could be constructed for some $9 billion and the Navy's avowed minimum goal of one such cruiser for each attack carrier achieved.[28] The money saved would be ample to finance the Tarawas and other vessels better suited to sealane protection and operations in the Third World.

The challenge of Soviet naval aviation can most effectively be met by western carrier airpower. Carrier based aircraft, both conventional and V/STOL, are the viable means of protecting the sealanes and supporting needed operations in the far corners of the world.

[1]Commander M. Forrest, "A Successor to the Sea Harrier," *Journal of the Royal United Services Institute for Defense Studies* (March, 1980), p. 38.

[2]*Ibid.*, p. 37.

[3]Anibal A. Tinajero, *Fleet Air Defense: A Naval Problem* (Washington, D.C.: Congressional Research Service, Library of Congress, September 19, 1979), pp. 70, 72.

[4]Jean Labayle-Couhat, ed., *Combat Fleets Of The World 1980/81: Their Ships, Aircraft, and Armament* (Annapolis, Maryland: The United States Naval Institute, 1980), pp. 630-636.

[5]The International Institute for Strategic Studies, *The Military Balance 1980-1981* (London: The International Institute for Strategic Studies, 1980), pp. 7-8.

[6]Norman Polmar, "The U.S. Navy: Naval Aircraft, Part 1," *Proceedings* (September, 1980), p. 121.

[7]Bill Gunston, *The Encyclopedia of World Airpower* (New York: Crescent Books, 1980), p. 186.

[8]British Aerospace, *The Crossroads For Naval Combat Aviation* (Kingston upon Thames, England: British Aerospace, 1979), p. 34.

[9]Angus Macpherson, "VIFF - The Agility Factor," *Air International* (December, 1974), pp. 263-267, 294.

[10]British Aerospace, *op. cit.*, p. 56.

[11]"Britain's V/STOL Navy," *Air International* (March, 1979), p. 114; and J.W. Fozard, *Ski-Jump: A Great Leap for Tactical Airpower* (Kingston upon Thames, England: British Aerospace, 1979), pp. 7-8.

[12]Roy M. Braybrook, "Military V/STOL . . . What Is The Future?" *Air Progress Aviation Review* (August, 1980), p. 60.

[13]*Ibid.*, pp. 59-60.

[14]Tinajero, *op. cit.*, pp. 95-96.

[15] Labayle-Couhat, op. cit., pp. 690-692.

[16] "Small Carrier & V/STOL Proposals Get Little Notice, No Funding," *Armed Forces Journal International* (April, 1980), p. 43.

[17] Paul Maurice, "Sea Harrier Status Report," *International Defense Review* (No. 6, 1980), pp. 865-866.

[18] Graham Warwick, "AV-8B Advanced Harrier," *Flight International* (December 29, 1979), pp. 2127-2142.

[19] Ibid., p. 2132.

[20] "UK puts off AV-8B/MK 5 Harrier choice," *International Defense Review* (No. 7, 1980), p. 983.

[21] Warwick, op. cit., p. 2142.

[22] Labayle-Couhat, op. cit., pp. 285-286.

[23] Ibid., p. 448.

[24] "Invincible Handed Over," *Defence* (March, 1980), pp. 150-152.

[25] R.B. Pengelley, "The Royal Navy's Invincible-class cruisers," *International Defense Review* (No. 8, 1979), pp. 1336-1338.

[26] Labayle-Couhat, op. cit., p. 690.

[27] Edmund J. Gannon and Alva M. Bowen, *Naval Shipbuilding Costs: A Projection* (Washington, D.C.: Congressional Research Service, Library of Congress, June 28, 1979), p. 37.

[28] Ibid., p. 31.

## CHAPTER IV
## Confronting the Soviet Submarine Threat

How can the United States and its western European allies cope with the very substantial and constantly expanding Soviet underseas fleet? Foxtrot and Victor class submarines are joined by the new and more advanced Tango and Alpha class warships. It is a formidable challenge indeed.

Reinforcement and resupply from North America to the European theatre is a fundamental element of western strategy. The sheer magnitude and crucial nature of this mission is underscored by Paul Nitze and associates:

> In the event of a NATO-Warsaw Pact war, the United States currently plans to move approximately 10 million tons of American war supplies and 15 million tons of fuel to NATO ports within the first ninety days of hostilities to reinforce U.S. forces. This operation would require roughly 1,000 merchant vessels.[1]

And Rear Admiral Sayre Swarztrauber notes:

> Moreover, resupply must include the minimum essential economic requirements of the NATO nations for food, fuel, raw materials, and manufactured goods to sustain the civilian population of some 300 million, the economies, and hence, the war effort.[2]

Failure to achieve these tasks could compel recourse to nuclear weaponry or abject surrender.

Moreover a war in Europe is unlikely to be an isolated event. Western naval forces may well be called upon to protect ships carrying oil and vital ores from the Persian Gulf and southern Africa on sea routes that are very long and quite vulnerable. Attack carrier task forces may be needed as conflicts erupt in other quarters of the globe. Shrunk in numbers and overcommitted, the American fleet might well find such conditions daunting.

Could the western alliance win a third Battle of the Atlantic? Charles DiBona and William O'Keefe state that: "Our combined NATO forces currently appear to have adequate levels of surface escorts to handle the initial requirements for the escorting of military cargoes across the Atlantic."[3] However, they carefully note that escorts are not available to protect the shipping of essential goods to western European economies. And escorts are not sufficient in numbers to assure protection on the long sealanes from the Persian Gulf. Nor are losses suffered by escort forces taken into account.

Many detailed studies have been made of the course a future war in the North Atlantic might take. Such forecasting, however, is extraordinarily difficult. As Charles DiBona and William O'Keefe point out:

> Comparisons of opposing capabilities should consider a number of variables, such as weapon characteristics and performance, warning, tactics, logistics, command and control, dynamics of conflict, and force size. . . . One of the major shortcomings of complex computer modeling is that it tends to assign a single performance value to each variable, even though there may be absolutely no test data to substantiate the point values chosen.[4]

What conclusions, then, can be drawn? From the welter of scenarios a range of possibilities emerge - none of which is reassuring. In the course of the first thirty days of war, one study indicates that half the merchant vessels might be sunk.[5] An authoritative recent analysis suggests that losses of shipping and escorts could range between 300 and 600 vessels in the initial stages of conflict.[6] And sorely needed weapons, munitions and fuel must be rushed to NATO forces battling desperately in western Europe. Under these circumstances a future Battle of the Atlantic might indeed be decisive.

How is the western alliance organized and equipped to wage an anti-submarine campaign? What are the essential elements in American strategy against the submarine? What part are western European navies prepared to play in defense of the sealanes? What is the current state of anti-submarine warfare?

Modern anti-submarine warfare is an extremely complex and technologically intricate matter. The means of submarine detection have been vastly improved. Passive sonar, particularly underwater fixed surveillance systems and towed arrays of hydrophones, can detect submarines at considerable distances. Indeed detection can now be made at ranges beyond the reach of ship-borne ASW weaponry. Increasingly, specific location and destruction of submarines is entrusted to helicopters and patrol aircraft. Much more is known about the ocean and the complex behavior of sound in water. Sound processing techniques, linked to computers, now make it possible to identify the unique noises of many kinds of submarines and differentiate such "signatures" from the noisy environment of the sea.

But modern submarines have improved substantially too. This is particularly true of nuclear fuelled submarines whose speed, range, and operating depth make them especially formidable weapons. And underwater fired missiles can be unleashed at a range beyond the ability of a surface warship to make a timely reply. Hence the submarine continues to be a deadly opponent.

The American fleet has developed the most varied and highly sophisticated anti-submarine technology and equipment. Much of the ASW weaponry in use by the western alliance, especially homing torpedoes, is American in origin and design. And 74 immensely capable nuclear fuelled attack submarines are in service with the U.S. Navy. However these submarines will be fully occupied countering their adversaries, Soviet SSBNs and SSNs, and protecting American attack carriers. Perhaps a few U.S. SSNs can be spared from these exacting tasks for sealane duty - but the main burden of such security must be the responsibility of other forces.

The anti-submarine strategy of the United States and its allies in the North Atlantic arena is based upon three interrelated and interdependent elements: the employment of convoying, a vast underwater sound surveillance system linked with a long range patrol aircraft, and intensively patrolled "barriers" in such key areas as the Greenland-Iceland-United Kingdom "Gap."

A modern convoy of merchant vessels would consist of some 60 ships escorted by six to eight ASW warships and perhaps two long range patrol planes.[8] Modern merchant ships are large, often more than 20,000 tons, and fast, capable of 20 knot speeds. The British, Dutch, and Canadian fleets play a crucial role - they furnish two thirds of the surface escorts needed to protect the North Atlantic convoys.

Convoying remains the most effective means of ensuring the delivery of vitally needed cargoes.[10] It continues to be an effective offensive tactic, compelling the submarine to fight on escort terms. And it makes possible the concentration of limited escort assets.

Far more convoy escorts would be needed, however, if it were not for the existence of SOSUS, the fixed underwater sound surveillance system. Hundreds of hydrophones are implanted on the ocean floor along the Atlantic coastlines of the United States and a number of western European nations.[11] Very long cables link the sonars to a number of stations on land. In these terminals sound processing equipment and computers sort submarine "signatures" out from the myriad noises of the sea. Specific location of a submarine however must be done by a land based, long range patrol plane - an essential partner in this system. Employing air dropped sonobuoys and magnetic anomaly detection gear a patrol plane can frequently pinpoint a submarine.

Some 200 Lockheed P-3 Orion patrol aircraft may be available for North Atlantic duties.[12] These turboprop engined machines have a maximum radius of 2,384 miles, feature an elaborate array of avionic and electronic systems, and are armed with eight homing torpedoes.[13] The Royal Air Force flies 28 of the excellent Nimrod turbofan powered ASW patrol and reconnaissance warplanes.[14] And the Canadians are taking delivery on a version of the Orion embodying the considerably advanced avionics suite of the Lockheed Viking S-3. Fourteen of these aircraft will be stationed on Canada's eastern seaboard.[15]

SOSUS is thus an invaluable source of information. It will enable long range ASW warplanes to inflict losses on Soviet submarines before they can close upon the convoys.

And it may be possible to "clear" large areas of the eastern approaches of the North Atlantic.

The SOSUS system does have its shortcomings, however, and it is vulnerable.[16] In wartime the long cables linking the hydrophone arrays to land stations can be slashed. The land terminals can be put out of action by air attacks or commando raids. In peace or war Soviet political pressure or nationalistic zeal can lead to the closing of stations located in foreign countries. A variety of air deployed systems are being explored and are under development. Such systems could be swiftly deployed and discreetly employed. Norman Friedman underlines the importance of SOSUS: "Any disaster to SOSUS would require a drastic change in U.S. forces and tactics."[17]

The third vital element of alliance strategy against the submarine is to be found in the "barriers." These defensive belts cover the ocean gaps through which Soviet submarines debouch upon the North Atlantic.

The most important of these "choke points" is the Greenland-Iceland-United Kingdom set of passages. To penetrate such defended zones Soviet submarines must run a gauntlet of fixed underwater hydrophones, minefields, allied "hunter-killer" submarines, and patrol aircraft. The strategic position of Iceland is absolutely crucial in a struggle in the North Atlantic. It is the "cork in the bottle." Iceland must be successfully defended against Soviet amphibious and airborne assault.

The significance of barrier operations is underscored by studies demonstrating that as much as 60 percent of Soviet submarine losses might be inflicted by these defensive systems.[18] Such barriers also make it possible for the western European allies to concentrate their patrol aircraft and diesel-electric "hunter-killer" submarines to best advantage. Operating under air cover and within defensive systems quiet conventional submarines can be quite effective in the ASW role.

Although much reliance is placed upon SOSUS, the patrol warplanes, and the barrier systems convoys remain essential to deliver the enormous tonnage of equipment, supplies, and fuel to the European theatre. Some 72 American frigates are available for escort purposes.[19] Not all of these warships could be committed to the North Atlantic, in a war many demands would be levied upon them.

Backbone of the current frigate fleet is the Knox class, 46 of which are in service. A 4,200 ton vessel at full load, the Knox features an anti-submarine rocket system which releases a homing torpedo and a helicopter.[20] A number of the Knox class have been equipped with Sea Sparrow anti-aircraft missile launchers. And Harpoon surface to surface missiles are also being retrofitted. Many of the Knox class carry variable depth sonars. Although bulky and expensive, active VDS can operate clear of the noise of the ship and can penetrate the different temperature layers of the water, unmasking "blind" zones where submarines may lurk.

A new class of patrol frigate, the Oliver Hazard Perry, is now entering service with the fleet. At the outset of the program a total procurement of 74 vessels was envisaged. But changes in the future configuration of the navy and rising costs may sharply reduce the planned total. It is difficult to predict how many Perrys will be completed. By the end of 1980 eight

Perrys should be operational.[21] Thus far 47 more units of the class are under construction or have been authorized.

Designed as a relatively less costly and easily supplied and maintained escort for convoys or amphibious assault forces, the Perry weighs 3,600 tons fully laden.[22] Powered by gas turbines the Perry can reach 30 knots, although maximum sustained speed is 28 knots. A roomy ship, the Perry can be swiftly replenished and damaged machinery and weaponry are readily accessible. Magazines and the engine control room are shielded by alloy armor. Electronic systems are protected by Kevlar plastic armor.

The Perry is to be armed with a launcher for medium range anti-aircraft missiles and Harpoon surface to surface rockets, a Vulcan Phalanx 20 mm. machine cannon for close defense, a rapid fire dual purpose 76 mm. cannon, six ASW torpedo tubes, and two ASW helicopters.[23] Unfortunately the vessels have not been furnished with their complete armament. The Perrys have yet to be outfitted with the gear needed to handle helicopters - their primary anti-submarine weapons. The new ASW helicopter system, LAMPS III - which the Perry is specifically designed to employ, will not become operational until 1984. The Vulcan system must also be retrofitted.

The frigates are incomplete in other respects as well. Provision is made for needed fin stabilizers which have yet to be fitted. Nor are the ships fully equipped with their planned electronic countermeasures suite.

A number of shortcomings mar the design of the Perry. The missile launcher occupies the forward deck - with a field of fire wider than needed - while the 76 mm. cannon, tucked away between a radar installation and the stack, consequently suffers a greatly constricted field of fire. A much more serious matter is that, in spite of much attention to protection and damage control, the command and control center of the Perry is located in the topside aluminum deckhouse - which is vulnerable to destruction by direct hits and by fire.[24] Even so, when fitted with all their equipment and the LAMPS III helicopters the Perrys should be versatile and formidable warships. Whether they will be secured in the numbers essential for sealane protection, however, may well be doubtful.

An extremely valuable addition to the anti-submarine effort is the TACTAS system or tactical towed acoustic sensor, a long line of hydrophones towed hundreds of yards behind the escort ship. The hydrophones are thus well clear of the noise of the vessel. And the escort can move at high speed. Moreover, as William D. Taylor points out:

> One of the submarine's major advantages is her ability to operate at the best search or tracking depth consistent with environmental conditions. Now, with towed systems the surface warship has gained a similar capability, and without sacrificing her ability to perform her other missions.[25]

Detection of submarines is thus possible at much greater distances. It may be possible to equip fast merchant vessels with this system. And older escorts might be modernized with towed line array sensors.

In order to destroy submarines at the longer ranges at which they can now be detected the helicopter is the indispensable partner. The ASW helicopter currently in widest service with American escorts is the Kaman SH-2F or Seasprite, equipped with search radar, sonobuoys, magnetic anomaly detection gear, and homing torpedoes. Helicopters are also essential in providing targeting information for surface to surface missiles such as Harpoon.

Eagerly awaited by the fleet, however is the new light airborne multi-purpose system III. The Sikorsky SH-60B or Seahawk features a maximum speed of 184 miles per hour, a range with 30 minute reserve of 373 miles, and the ability to function at night and under adverse weather conditions.[26] The heart of the system, however, is an advanced digital data link and sophisticated signal processors which enable shipboard operators to instantaneously communicate and control the helicopter in its pursuit of a submarine.[27] Thus ship and helicopter act as one integrated weapon system. LAMPS III is scheduled to enter the fleet in 1984 and will be the key ASW weapon for the Perry class frigate and the Spruance class destroyer. Other warships, lacking the space for the elaborate onboard electronic suite or the larger helicopter, will continue to operate the Seasprite.

What contribution can other NATO navies make to the security of the North Atlantic sealanes? Canada, France, Great Britain, the Netherlands, and West Germany maintain deep water fleets. The British, Canadian, and Dutch navies play a crucial role in the Atlantic. The French and West German fleets are heavily committed to other theatres.

France musters a respectable, balanced, and versatile conventional fleet. It includes two medium size attack carriers, a command cruiser, 18 destroyers, and 21 diesel-electric, "hunter-killer," submarines.[28] The carriers have undergone thorough refit and modernization. Among the new warships are the heavily armed and powerful Tourville class destroyers.[29] And eight of the excellent Georges Leygues class ASW and AA vessels are under construction. Variable depth sonar is emphasized in ASW operations. French conventional submarines are very quiet, deep diving, and maneuverable anti-submarine weapons. A modest capability for amphibious assault exists, centered around a helicopter cruiser and two specialized transports. Countermeasures against mines receive attention; some 35 minehunters and minesweepers, including five vessels of recent construction, are in service.

The French have been slow to employ gas turbine engines in their warships and to bring ASW helicopters into operation. Eleven of the destroyers are of 1950s vintage and, although repeatedly updated, will come to the end of their useful life in the 1980s. And ten of the submarines are some 20 years old and will need to be retired.

French naval forces do work frequently with NATO fleets in exercises and maneuvers. However French naval strategy is focused upon the Mediterranean and the Indian Ocean. The British have concentrated their naval resources in the Channel and the eastern Atlantic. And the United States Navy is overcommitted and severely strained. "It is in recognition of this," Bernard Ireland points out, "that France has moved the major strength of her fleet to its southern base of Toulon."[30]

One hardly needs to belabor the significance of the

Indian Ocean in the light of current events in that violently unstable region. Based on Djibouti and the island of Réunion the French maintain a naval presence in the area. From time to time the French have sent a carrier task force into the Indian Ocean.

The West German Navy is a small but efficient and highly disciplined service. Organized in the main for operations in coastal waters it is equipped with 30 fast, heavily armed guided missile patrol craft, 24 small submarines, and 56 mine warfare vessels.[31] Six light frigates and five corvettes configured for ASW operations in shallow waters are also to be found in the West German naval array. But these vessels date from the early 1960s and their armament is obsolete. Under construction are ten more swift missile boats to be completed in the early 1980s.

The West German fleet does possess some ten deep water combatants however. Three destroyers are modified versions of the American Adams class. Four destroyer units are the German designed and built Hamburg class, recently furnished with French Exocet surface to surface missiles. But three more units are ancient American Fletchers from the Second World War era.

Fortunately, six well armed Type 122 general purpose frigates will be built in the early 1980s to supplant the aging Fletchers and Cologne class frigates.[32] These vessels are designed to meet the German Navy's need for a ". . . general-purpose frigate with long-range ASW and anti-surface armament and a respectable short and very short-range air-defense capability."[33] When completed the new Bremen class warships should fulfill these demands admirably.

The West German fleet is organized, equipped, and trained primarily for defensive operations in the Baltic and North Seas. As Soviet naval and amphibious strength in the Baltic has increased and Scandinavian capabilities have suffered relative decline, West German responsibility for defense of these vital regions has grown much heavier.

Developments in the spring and summer of 1980, however, have complicated the strategic situation for the West German Navy. The crises in the Indian Ocean have levied heavy demands upon the naval resources of the United States. And the British fleet has dispatched units to that troubled and vital area. These circumstances raise the possibility that the West Germans may be called upon to patrol in the Norwegian Sea. In June 1980, the German Federal Security Council lifted the ban upon naval operations more than a day's sailing from the straits leading into the Baltic sea.[34] The following month the Council of the Western European Union removed restrictions, dating from West German rearmament in 1954, on West German naval construction.

But the West German fleet is ill-equipped to play a deep water role. Only the Adams and Hamburg class warships are suitable for the mission. Unfortunately, the West German government is in no position to remedy this situation. Military expenditures for German air arms and Army weaponry are very heavy, particularly the Tornado warplane and Leopard II tank programs. Moreover the full impact of recession has now hit hard: an unfavorable balance of trade, rising unemployment, and sharply plummeting tax revenues. Budget stringencies are now forcing the West German government to limit Tornado procurement and stretch out production.[35] And to cap the matter the warships deemed essential by the WEU for operations in northern waters are heavy destroyers - 6,000 ton vessels.[36]

Under all these circumstances the West German Navy will be hard pressed to sustain a viable presence in the Norwegian Sea. If the western alliance deems an immediate expansion of West German naval operations imperative then it may be necessary to loan appropriate warships from the U.S. Navy. Under such an arrangement, the West Germans would crew the ships and pay their operating expenses. The Norwegian Sea is a lethal combat environment. To be effective in this area, a warship must possess formidable anti-aircraft and surface to surface weaponry. Just such vessels are to be found in the four Kidd class heavy destroyers, completed to an Iranian order and subsequently cancelled by the revolutionary regime. Standard missile and Vulcan-Phalanx gun anti-aircraft systems, Harpoon rocket and five inch gun armament, and ASROC and Sea King helicopter can all be carried by the Kidd.[37] The venerable Fletchers and aging Colognes could be speedily retired and their crews transferred to modern ships.

The Canadian Maritime Command is also a highly specialized fleet. However Canada logically concentrates on deep water operations in the North Atlantic. The Canadians have earned a justly merited reputation for anti-submarine expertise in the two world wars. Warships especially suited for icy northern waters are designed and built in Canada.

Stationed on the Atlantic seaboard are four heavy destroyers and twelve frigates.[38] The Canadians employ the large and very well equipped Sea King ASW helicopter. Four obsolete frigates are employed in the training role. Especially impressive are the Iroquois class destroyers.[39] Powered by gas turbine engines, the Iroquois is fast and far ranging. It is armed with Sea Sparrow anti-aircraft missiles, a five inch dual purpose cannon, ASW torpedo tubes, carries two Sea King helicopters, and is equipped with variable depth sonar.

But at least eight outdated Mackenzie and Restigouche class frigates must be replaced. Designs are now under examination for six new frigates to be built in the course of the 1980s. Armament of the proposed frigate will include a new anti-aircraft missile system, a 76 mm. dual purpose gun, surface to surface missiles, and a Sea King. The new warship will be furnished with towed array sonar.

Another excellent deep water fleet is the Royal Netherlands Navy. Bearers of a proud naval tradition, the Dutch maintain a substantial fleet. Two fine new command destroyer leaders, nine thoroughly modernized or new frigates, and five older destroyers are in service.[41] Nine more frigates are under construction. Six diesel-electric ocean going submarines are operational and two new craft are building. And thirteen of the long range and very effective Lockheed P-3C Orion patrol aircraft are on order.

Moreover the Dutch Navy is primarily committed to anti-submarine operations in the North Atlantic. The fundamental basis of Dutch naval strategy is succinctly outlined by Captain F. deBlocq van Kuffeler:

> The surface forces will include three task groups, each consisting of a flagship/air-defense destroyer, six anti-submarine frigates and a fast fleet supply ship for operations in the Atlantic and the Western Approaches. . . . Therefore, in the early 1970s, two large guided-missile destroyers of the Tromp class were built. These are now being joined by twelve 3,500-t Standard Frigates. . . . At the same time, the Van Speijk class frigates. . . . are being given a mid-life modernisation. A third flagship/air-defense vessel will be ordered soon. This program is due to be completed by about 1983.[42]

The Standard or Kortenaer class frigate may well be the most efficient and best balanced design of all the new escort warships coming in service.

Weighing 3,750 tons fully laden, the Kortenaer is powered by gas turbine engines and can reach a maximum of 30 knots.[43] Armament is well balanced and comprehensive. Eight Harpoon surface to surface missiles are included, as is a 76 mm. dual purpose cannon. Anti-aircraft weaponry comprises Sea Sparrow missiles and a quadruple unit of 30 mm. machine cannon. Two Lynx helicopters bear the main burden of anti-submarine operations, albeit four ASW torpedo tubes are fitted for short range work.

Command and control rooms are located in the hull where they can be best protected. This "citadel" can be sealed and pressurized against nuclear, chemical, or biological attack. The main systems of the vessel can be operated by remote control.

A high degree of automation and computerization is incorporated in the Kortenaer.[44] It is thus possible to handle the ship effectively with a minimum of crew.

As much of the armament and machinery as possible

was chosen with an eye to standardization within the NATO alliance. The Harpoon and Sea Sparrow missiles are American, the Lynx is Anglo-French, and the gas turbines are English in origin.

The hull is designed in such a way that a wide variety of different armaments, electronic suites, and propulsion machinery can be fitted. Hence the initial designation of the Kortenaer as the Standard class. The West German Type 122 or Bremen class frigates now building employ the Dutch design with different sonar and engines.

The major deep water fleet in western Europe, however, is that of Great Britain. Eleven nuclear fuelled "hunter-killer" submarines are operational and a twelfth will be accepted by the end of 1980.[45] Three more nuclear attack submarines will be on order. Also in service are sixteen ocean going diesel-electric submarines and a new conventional patrol submarine is being designed. Two V/STOL carriers are in service and two more are under construction. Some dozen guided missile destroyers are to be found in the air defense role; seven more have been laid down. Anti-submarine frigates in active service number approximately 45.[46] And four of the new Broadsword class are building or on order. A commando carrier, two assault ships, and six specialized cargo vessels support a fair amphibious capability. Thirty seven vessels are dedicated to mine countermeasures, including a new class of minehunter with a glass reinforced plastic hull.

The significance of the Royal Navy is underlined by the fact that 80 percent of the ships in the eastern Atlantic command are contributed by the British fleet.[47] Logically, English admirals hold the key posts of Commander-in-Chief Channel and Commander-in-Chief Atlantic.

As befits a major maritime trading nation, the Royal Navy is organized, equipped, and trained for anti-submarine warfare. British naval forces are, however, quite capable of global operations as well. The new V/STOL carriers will restore balance to the fleet and provide organic air for viable task force operations in distant waters.

Unfortunately, the chronic difficulties of the British economy have had a severe and constant impact upon the armed forces. Ironically, some benefits have accrued from certain economic pressures. The Royal Navy is, in the main, a modern fleet. Warship construction programs have been driven in no small part by a concern to keep jobs in an industry and in regions suffering high unemployment. But unhappily the handicaps imposed by economic stringencies far outweigh the few advantages. A case in point is painfully to be seen in present circumstances. Reeling under the double blows of massive depression and high inflation, the government must confront the harsh necessity of slashing the defense budget for the fiscal year beginning in April of 1981 by $480 million. And this in spite of the conservative government's strong sympathy and support for the armed forces. The future consequences for the British fleet are clearly put by David Brown:

> The Royal Navy had been told earlier that the funds for the Trident (successor to the current Polaris system in British SLBM force) would not be charged against its share of the defense budget, but would come from a special strategic section in the ministry's overall budget. Now the service has been told that the money will have to come from its budget after all, which senior officials have complained will severely limit the ability of the service to modernize its antisubmarine warfare capability,

which it sees as its top priority.[48]

In spite of many difficulties, the Royal Navy still is the largest and, in many respects, the most capable western European fleet. Several features of British warship design need to be emphasized. English vessels are built to fight effectively in the harsh and demanding conditions of the North Atlantic. Hence seaworthiness is stressed, the ability to function in heavy seas and rough weather.

A certain measure of protection is also a key priority in British design. Most modern navies employ aluminum in warship superstructures. Aluminum reduces weight and enhances stability. But unfortunately aluminum is inflammable and when hit can disintegrate in a hail of deadly splinters. The British have minimized the use of aluminum and employ steel in warship construction. Consequently, as Captain J.W. Kehoe and associates note:

> The British Type 42, with little ordnance exposed topside, a steel superstructure with heavy scantlings, adequate mechanical and electrical redundancy, and CIC and computer room located in the hull, is well designed to resist the effects of small conventional weapons and fragments.[49]

English vessels can seal and pressurize vital compartments against nuclear fall-out, chemical attack, or biological agents. A price is exacted for such steel protection however. As Norman Friedman indicates: ". . . the price accepted for a more survivable hull was a lesser weapon and sensor suit on the same dimensions or, conversely, a larger hull for a given suit."[50] This problem is exacerbated by the bulkiness and heavy weight of some key British weapon systems and electronic equipment.

Backbone of the British ASW force is the Leander class frigate. Twenty six of these vessels are in service with the Royal Navy, five of which are currently undergoing comprehensive modernization. Leander is a very successful warship with a wide appeal. Versions of Leander have been purchased or built under license for the navies of Australia, Chile, India, the Netherlands, and New Zealand.

In Royal Navy service Leanders fall into three main types.[51] All range in weight, fully loaded, between 2,860 and 3,200 tons; and are capable of 27 knots. And all are furnished with an ASW Wasp or anti-surface vessel Lynx helicopter.

One group of Leanders is armed with Ikara, a guided missile, carrying a homing torpedo, which has a maximum practical range of twelve nautical miles. It can be given terminal guidance by the Wasp ASW helicopter. Variable depth sonar is fitted to provide long range detection capability. Two short range Sea Cat rocket launchers give protection against close in air attack.

The second group of Leanders now has four Exocet surface to surface missiles and three Sea Cat AA rocket launchers. Six ASW torpedo tubes provide a short range capability.

A third group, whose modernization is now in hand, will be armed with four Exocet missiles and enlarged facilities for the Lynx helicopter. Lynx will be equipped with air to surface missiles as well as anti-submarine torpedoes. And the Sea Wolf anti-missile defensive system will be installed.

Ships drawn from each of the three groups can thus

form a balanced and respectable task force. Such a force will possess an offensive capability against submarines and surface warships and a defensive ability against air and missile attack.

A new ASW frigate is the Type 22 or Broadsword. Two of these vessels are in service and four more are building or on order.[52] At 4,000 tons fully loaded the Broadsword, powered by gas turbine engines, can make 30 knots.[53] Main armament comprises two Lynx helicopters, four Exocet missiles, and two Sea Wolf multiple launchers. Six ASW torpedo tubes are also fitted.

Broadsword incorporates a new multiple frequency sonar linked to a computer. Space is available for a towed array sonar as well. Communications systems are comprehensive, embracing all frequencies. Satellite communications terminals can also be installed. Command and control are assisted by a computer information, data processing, and weapon automation system.

Broadsword will be an integral part of a team built around the new Invincible carriers. Sheffield class destroyers will provide medium range anti-aircraft missile cover. The Broadswords will be entrusted with anti-missile and close in anti-submarine defense. "According to the Royal Navy," R.B. Pengelley indicates, "such task groups are likely to operate in areas such as the Greenland/Iceland/UK gap . . . or in direct support of convoys shipping NATO reinforcements from North America to Western Europe."[54]

Although some 37 vessels are devoted to defensive minesweeping and minehunting mine countermeasures are sharply limited by financial considerations.[55] However British capabilities will be enhanced as the new Hunt class of multiple purpose mine countermeasures vessel comes into service. Brecon, the first of the class, is operational and three more are building or on order. The Royal Navy would like to secure at least a dozen of these advanced minehunters.56 But such intentions may be dimmed by the high cost of the new warship.

Brecon is an innovative and sophisticated ship.57 It employs a glass reinforced plastic hull to reduce magnetic attractions to mines to a minimum. Seakeeping is designed to cope with the stormy North Sea and western approaches. A wide variety of sweeping and hunting systems are carried. In addition to sonar, three navigational systems are linked to a central computer system to achieve precision location and efficient sweeping. The Hunt class should prove to be versatile ships, suitable for a wide range of patrol duties as well as mine countermeasures.

What can be done to strengthen the protection of convoys running the gauntlet of Soviet submarines? Western European and Canadian fleets do make a crucial contribution to convoy escort in the North Atlantic. American and western European anti-submarine operations are complementary and mutually reinforce each other.

What is clearly needed is a larger force of escort frigates. Such warships should be versatile and capable vessels. Money devoted to light ships limited to a single role is ill spent.

It should be possible to form a number of escort squadrons, drawn from the alliance navies, patterned upon the highly successful Standing Naval Force Atlantic. Such a force might be funded as a major NATO project.

Perhaps alliance sealane security squadrons could be equipped with a standard escort frigate. Substantial warship surplus building capacity is to be found in British and, by 1983, Dutch shipyards.[58] Such English builders as Yarrow and Cammell Laird have modernized their yards and construction facilities. Dutch yards are very well organized and, computer controlled, quite efficient.[59]

Two excellent possibilities exist for an alliance standard frigate. Yarrow of Great Britain has developed a design for a general purpose frigate, the Type 24.[60] A wide variety of weapon systems and sensors can be fitted to meet different roles. American, British, French or Italian weapons and electronic systems can equally readily be accommodated. Maximum flexibility is thus a key feature of the 3,100 ton design.

Another fine candidate for an alliance escort - and one which is in operational service - is the Dutch Kortenaer. The new Bremen class, building in West German yards, is based upon the Dutch design. And the Kortenaer is a leading contender in the new Canadian frigate program. Both the Kortenaer and the Type 24 design are highly automated and consequently fewer men are needed to operate these vessels.

An alliance escort frigate force could be a very effective representative of the transatlantic community and a most appropriate guardian of the sealanes that knit that community together.

¹Paul H. Nitze, Leonard Sullivan, Jr., and the Atlantic Council Working Group on Securing the Seas, *Securing the Seas: The Soviet Naval Challenge and Western Alliance Options* (Boulder, Colorado: Westview Press, 1979), p. 380.

²Rear Admiral Sayre A. Swarztrauber, "The Potential Battle of the Atlantic," in Frank Uhlig, Jr., ed., *Naval Review 1979* (Annapolis, Maryland: U.S. Naval Institute, 1979), pp. 115-116.

³Nitze, et. al., *op. cit.*, pp. 347, 349.

⁴*Ibid.*, p. 369.

⁵*Ibid.*, p. 345.

⁶*Ibid.*, p. 374.

⁷The International Institute for Strategic Studies, *The Military Balance 1980 - 1981* (London: The International Institute for Strategic Studies, 1980), p. 7.

⁸Nitze, et. al., *op. cit.*, pp. 345-346.

⁹*Ibid.*, pp. 201, 203-204.

¹⁰Commander William F. Mellin, "To Convoy Or Not To Convoy," *Proceedings: United States Naval Institute* (March, 1980), pp. 50-52; and Captain William D. Taylor, "Surface Warships Against Submarines," in Frank Uhlig, Jr., ed., *op. cit.*, pp. 179-180.

¹¹Norman Friedman, "SOSUS and U.S. ASW Tactics," *Proceedings: United States Naval Institute* (March, 1980), pp. 120-123.

¹²Nitze, et. al., *op. cit.*, p. 347.

¹³Bill Gunston, Consultant Editor, *The Encyclopedia of World Airpower* (New York: Crescent Books, 1980), pp. 231-232.

[14]The International Institute for Strategic Studies, op. cit., p. 23.

[15]Lieutenant Maurice S. Joyce, "Dawn of Aurora," Proceedings: United States Naval Institute (March, 1980), pp. 127-129.

[16]Friedman, op. cit., p. 120.

[17]Ibid., p. 123.

[18]Nitze, et. al., op. cit., p. 360.

[19]The International Institute for Strategic Studies, op. cit., p. 7.

[20]Jean Labayle-Couhat, ed., Combat Fleets Of The World 1980/81: Their Ships, Aircraft, and Armament (Annapolis, Maryland: The United States Naval Institute, 1980) pp. 681-684.

[21]Ibid., p. 678.

[22]Ibid., p. 680.

[23]Paolo Penoni, "Oliver Hazard Perry," Aviation and Marine International (June, 1978), pp. 53, 56-58.

[24]Captain J. W. Kehoe, Commander C. Graham, K.S. Brower, and H. A. Meier, "NATO and Soviet Naval Design Practice - eight frigates compared," International Defense Review (No. 7, 1980), pp. 1005-1006.

[25]Taylor, op. cit., p. 179.

[26]Gunston, op. cit., pp. 326-327.

[27]Rear Admiral Raymond N. Winkel and Dan Manningham, "LAMPS: The Ship System with Wings," Proceedings: United States Naval Institute (March, 1980), pp. 114-117.

[28]Labayle-Couhat, op. cit., pp. 98-99.

[29]Ibid., pp. 123-124.

[30] Bernard Ireland, "The French Navy Tody," *Armed Forces* (No. 7, 1980), p. 22.

[31] Labayle-Couhat, op. cit., pp. 174-180.

[32] Udo Philipp, "The German Navy's Type 122 General-purpose Frigate," *International Defense Review* (No. 4, 1978), pp. 539-544.

[33] Ibid., p. 540.

[34] Thomas Ries, "Expanding Horizons for the Bundesmarine," *International Defense Review* (No. 7, 1980), p. 986.

[35] "Germans Seek Tornado Production Stretch," *Aviation Week and Space Technology* (November 10, 1980), p. 23.

[36] Ries, op. cit.

[37] Labayle-Couhat, op. cit., p. 668.

[38] Ibid., pp. 43, 46.

[39] Ibid., p. 45.

[40] Kehoe, et. al., pp. 1004-1005.

[41] The International Institute for Strategic Studies, op. cit., p. 30.

[42] Captain F. de Blocq van Kuffeler, "The M-class Frigate," *International Defense Review* (No. 4, 1980), p. 504.

[43] Labayle-Couhat, op. cit., p. 373.

[44] Captain F. de Blocq van Kuffeler, "The Royal Netherlands Navy's Standard Frigate," *International Defense Review* (No. 4, 1978), pp. 522-523.

[45] Cmnd. 7826-1, *Statement on the Defence Estimates 1980, Volume I* (London: Her Majesty's Stationery Office), pp. 72-73.

[46] Labayle-Couhat, op. cit., pp. 208-215, and Addenda.

[47]House of Commons, *Minutes Of Evidence Taken Before The Defence Committee: Maritime Environment* (London: Wednesday 16 April 1980), p. 41.

[48]David A. Brown, "Cuts Imperil British Force Planning," *Aviation Week and Space Technology* (November 10, 1980), p. 57.

[49]Kehoe et. al., op. cit., p. 1006.

[50]Norman Friedman, *Modern Warship: Design and Development* (Greenwich: Conway Maritime Press, 1979), p. 169.

[51]Labayle-Couhat, op. cit., pp. 211-212.

[52]R. B. Pengelley, "The Royal Navy's Type 22 Frigate," *International Defense Review* (No. 1, 1980), p. 74.

[53]Ibid., p. 73.

[54]Ibid., p. 74.

[55]House of Commons, op. cit., pp. 53-55.

[56]"MCM proposals-a multitude of options," *International Defense Review* (No. 9, 1979), p. 1561.

[57]R. B. Pengelley, "The Royal Navy's New Mine Countermeasures Vessel," *International Defense Review* (No. 1, 1979), pp. 84-88.

[58]Van Kuffeler, "The M-class Frigate," op. cit., p. 506; and R. J. Daniel, "Warship Building by British Shipbuilders," *Journal of the Royal United Services Institute for Defense Studies* (March, 1980), pp. 73, 75.

[59]Van Kuffeler, "The Royal Netherlands Navy's Standard Frigate," op. cit., p. 524.

[60]R. W. S. Easton, "Britain's New Type 24 Multi-Purpose Frigate," *International Defense Review* (No. 6, 1979), pp. 1021-1025.

## CONCLUSION

The measures needed to protect the sea and air lanes across the North Atlantic fall into two categories. Some key situations demand remedy as soon as possible. Other problems can be met by long range policies.

Until Tornado interceptors come into service with the Royal Air Force immediate steps ought to be taken to buttress the air defense of Great Britain. Older Hunter fighters in some numbers should be pressed into service by the English to defend seaports, naval bases, and airfields. Several squadrons of United States Navy F-14 fighters and supporting aircraft should be stationed in northeastern Scotland to thwart Soviet Backfire bomber attack. Another United States Air Force tanker squadron based in England, is needed to augment the defensive radius and flexibility of British and American interceptors. Until Nimrod AEW becomes operational with the Royal Air Force, an American Sentry AWACS should be assigned to Britain.

If American attack carriers are to be a viable force F-14 fighter strength must be substantially and speedily increased. And units flying the long range, all weather A-6E must be restored to full strength - indeed new squadrons should swell their ranks.

Fortunately these measures can be swiftly implemented. Given a top priority, Grumman can increase F-14 production to 60 in Fiscal Year 1982 and 96 in the 1983 Fiscal Year, and turn out 18 A-6Es and EA-6Bs a year commencing Fiscal Year 1981.[1]

In the longer term, a considerable force of V/STOL

fighter bombers should be developed and brought into service. There are simply too many vital areas in the world - from northern Norway to the Indian Ocean - where fixed air bases and long runways may be too vulnerable or local political regimes too volatile and unstable. An improved version of the AV-8B should embody a more powerful engine and a suitable radar. Such warplanes can be readily accommodated by the current fleet of Iwo Jima and Tarawa class vessels. Perhaps the older carrier Oriskany can be overhauled and converted to a V/STOL carrier. It may be desirable to build four more Tarawas dedicated to V/STOL operations.

Strong support is to be found in the armed forces and in Congress for a V/STOL force at sea. The Marine Corps has repeatedly indicated its need for 336 AV-8Bs. In November, 1980, the Senate Appropriations Committee voted 333 million dollars for the development and initial procurement of the AV-8B.[2]

The development of a strong ASW frigate force devoted to North Atlantic sealane patrol is another major long term project. Such a force would be an expansion of the highly successful multi-national Standing Naval Force Atlantic. As is the case in STANAVFORLANT, the alliance force would include vessels from each nation within its squadrons.

It may be possible to secure a common ASW frigate for the North Atlantic navies. The Dutch Kortenaer is an excellent, proven ASW frigate. In the Type 24 the English have a design that might be developed into a suitable vessel. Key British warship building yards have been modernized and highly efficient production facilities are to be found in the Netherlands.

An alliance ASW frigate force needs medium range anti-aircraft missile protection. Just such a contribution could be made by the United States - assigning to each ASW squadron a Perry class frigate armed with the Standard missile system.

Where is the money to be found to implement these measures? Both Congress and the Reagan administration are increasing military expenditure sharply. But a substantial reduction in taxes has been promised. And in order to fight inflation the federal budget must be brought into balance as much as possible. To be sure reductions in non-military federal programs will be considerable. Even so, reconciling the imperatives of military spending, tax reduction, and balanced federal budgets will prove to be a challenge. And, even within the military budget, pay to keep highly trained people and ammunition, spare parts, and maintenance needed for operational readiness must receive overdue high priorities.

What is to be done? The most expensive weapon program in the United States Navy is the F-18 fighter and light attack plane, now in the process of development. But the F-14 in current service is a far better fighter than the F-18, superior in speed, radar, and missile armament.[3] And recent experience in the Indian Ocean calls the light attack role into serious question. A premium is now placed upon long range and all weather capability - the very qualities in which the present A-6E excels.[4] Cancellation of the F-18 would make available 30 billion dollars in future military spending.

Another costly system under development is the Aegis

anti-aircraft radar and missile system. But the most effective defense of the fleet against air attack is the F-14 and its improved Phoenix missile system. And a reduction in the number of Aegis cruisers to be procured could secure far more F-14s to defend the attack carrier task forces. Ten billion dollars in future military expenditure could be devoted to other needs - if Aegis construction were concluded with twelve cruisers.

Many more badly needed F-14s and A-6Es could thus be secured. The F-14 would become the key fighter of the fleet. Funds would be available to develope an improved AV-8B force and the carriers to support them. The alliance ASW frigate force could also draw substantially upon such monies.

Protection of the North Atlantic lines of communication and transport can only be achieved through the close cooperation of the western alliance. The United States, Canada, Britain, and the nations of western Europe constitute a maritime community, drawn closer together by growing economic ties. Even more important, the peoples of the North Atlantic hold in common a broad range of fundamental political values. These interests of the transatlantic community can be well served by a cooperative enterprise, a powerful and effective alliance ASW force.

¹Clarence A. Robinson, Jr., "Reagan to Accelerate Acquisition," <u>Aviation Week and Space Technology</u> (November 24, 1980), p. 18.

²Alton K. Marsh, "Senate Unit Boosts Funds for Defense," <u>Aviation Week and Space Technology</u> (November 24, 1980), p. 19.

³Anibal A. Tinajero, <u>Fleet Air Defense: A Naval Problem</u> (Washington, D.C.: Congressional Research Service, Library of Congress, September 19, 1979), pp. 39, 41.

⁴"Brown Slows F-18 by 40% in FY82-82," <u>Armed Forces Journal International</u> (September, 1980), pp. 17, 22.